Annals of Mathematics Studies

Number 110

THREE-DIMENSIONAL LINK THEORY AND INVARIANTS OF PLANE CURVE SINGULARITIES

BY
DAVID EISENBUD
AND
WALTER NEUMANN

PRINCETON UNIVERSITY PRESS

Princeton, New Jersey

1985

Clothbound editions of Princeton University Press books are printed on acid-free paper, and binding materials are chosen for strength and durability. Paperbacks, while satisfactory for personal collections, are not usually suitable for library rebinding

ISBN 0-691-08380-0 (cloth)
ISBN 0-691-08381-9 (paper)

Printed in the United States of America
by Princeton University Press, 41 William Street
Princeton, New Jersey

☆

Library of Congress Cataloging in Publication data will
be found on the last printed page of this book

CONTENTS

ABSTRACT

This book gives a new foundation for the theory of links in 3-space modelled on the modern development by Jaco, Shalen, Johannson, Thurston, et al., of the theory of 3-manifolds. The basic construction is a method of obtaining any link by ''splicing'' links of the simplest kinds, namely those whose exteriors are Seifert fibered or hyperbolic. This approach to link theory is particularly attractive since most invariants of links are additive under splicing.

Specially distinguished from this viewpoint is the class of links, none of whose splice components is hyperbolic. It includes all links constructed by cabling and connected sums, in particular all links of singularities of complex plane curves. One of the main contributions of this monograph is to the calculation of invariants of these classes of links, such as the Alexander polynomials, monodromy, and Seifert forms.

Three-Dimensional Link Theory and Invariants of Plane Curve Singularities

INTRODUCTION

In the first part of this introduction we will try to describe, loosely and informally, the problems and examples that led to this monograph, their historical setting, and the extent of our success. In the second part we will describe in more specific and technical detail the results achieved. The reader with an aversion to hot air, and the expert in the field, may skip part 1 without loss.

Review

In the last decade three-manifold theory has changed dramatically. We now picture a three-manifold (compact and with boundary, say) as decomposing canonically into pieces, each of which has a geometric structure in roughly the same sense that closed orientable two-manifolds "are" Riemann surfaces. This view is known to hold for most three-manifolds and is conjectured for all [Th 1]. The pieces of which the three-manifolds seem to be made come in two essentially different types: those which admit hyperbolic structures, and those which are or nearly are fibered over surfaces, the key examples being Seifert-fibered manifolds, which are circle bundles over surfaces except for finitely many degenerate fibers of a certain sort.

Among the most interesting example of three-manifolds have always been the exteriors of classical knots and links, made by removing from the three-sphere S^3 a disjoint union of open tubular neighborhoods of embedded circles. The geometric picture above is known to be correct in this case.

In the first part of this monograph we will show how the view of three-manifold theory loosely enunciated above can be applied to knot theory;

the key observations are that the pieces that arise are themselves associated to other links, so that a knot or link can be constructed canonically by pasting together irreducible pieces in a new sense; that the pieces that arise are all either hyperbolic or Seifert fibered; and that a large number of the classical invariants are essentially additive for this kind of decomposition.

Of course the idea of computing invariants of knots by additivity over some sort of decomposition is hardly new: the computation of the Alexander polynomial of a cabled knot from that of the uncabled knot and the cabling data by Torres [To] is a typical example of what has been done (and happens to be a special case of our results). The virtue of our treatment is that the decomposition with which we work is so fundamental and intrinsic.

The original motivation for our work was not, however, to give a new foundation for the theory of links. Rather we wished to give a "good" computation for certain of the invariants of the links arising from singularities of complex plane curves. To explain this context and how it developed, we will describe some of the history of these objects.

Let $f(x,y)$ be a complex polynomial in 2 variables, and consider the curve $f(x,y) = 0$. In the neighborhood of a smooth point of this curve the equation $f(x,y) = 0$ can be solved for one of the variables in terms of the other, but in the neighborhood of a singular point this is not so.

Newton [N] seems to have been the first to successfully investigate this matter, and he showed that in the neighborhood of any point one can give approximate solutions for one variable in terms of a root of the other variable. He actually gave a method for finding successive approximations of this type, involving the now famous Newton polygon; see for example [W] for a simple description. Taking the point of interest to be $(0,0)$, and supposing that the tangents to the curve are not in the direction of the y-axis, the method leads to a sequence of approximations of the form:

$$y_1 = a_1 x^{q_1/p_1}$$

$$y_2 = x^{q_1/p_1}(a_1 + a_2 x^{q_2/p_1 p_2})$$

$$y_3 = x^{q_1/p_1}(a_1 + x^{q_2/p_1 p_2}(a_2 + a_3 x^{q_3/p_1 p_2 p_3}))$$

$$\vdots$$

with p_i, q_i relatively prime and > 0.

Of course $f(x,y) = 0$ may have several solutions of this type near $x = y = 0$, corresponding to the different branches of the curve through $(0,0)$,

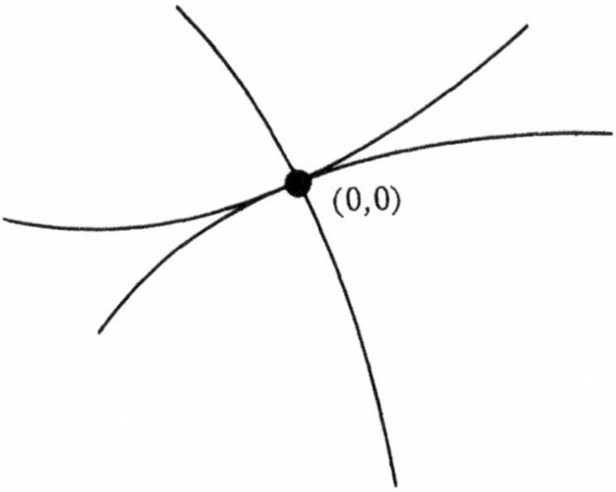

3 branches through (0,0)

and Newton's method finds them all.

In the middle of the nineteenth century Puiseux wrote these solutions in the more familiar form of fractional power series, or "Puiseux series,"

$$y = b_1 x^{m_1/n_1} + b_2 x^{m_2/n_1 n_2} + b_3 x^{m_3/n_1 n_2 n_3} + \cdots$$

with m_i, n_i relatively prime, and proved convergence. Since that time the singularities of plane curves have been identified by their "Puiseux pairs" $(m_1,n_1),(m_2,n_2),\cdots$, or, more recently, by a certain finite subset of this called the set of "characteristic pairs," which carries all the topological information. (In fact the equivalent finite subset of the set $(p_1,q_1),(p_2,q_2),\cdots$, which might reasonably be called the set of "Newton pairs" is more convenient for many purposes. For example, only the first Newton pair changes when one blows up the origin in the plane in the resolution process.)

The topology of the complement of the curve $f(x,y) = 0$ in a neighborhood of a singular point was first studied to understand ramification of complex surfaces. Already in the second half of the nineteenth century curves were often studied as coverings of the Riemann sphere. Such coverings ramify only over isolated points, and a huge role in the theory is played by the fact that the complement of a point in a small neighborhood in the plane has cyclic fundamental group; indeed, this is why it is enough to introduce roots of variables to solve $f(x,y) = 0$, as above. Around the turn of the century the theory of complex surfaces (real dimension 4) was rapidly developing. These surfaces can be represented as finite branched covers of the projective plane, branched along a curve. The converse of this statement – the analogue for P^2 of the "Riemann existence theorem" – was proven by Enriques [En] in 1923. The first step in the study of these branched covers (which is still an active subject), or, in a special case, of solutions to an algebraic equation $g(x,y,z) = 0$, was to understand the branching locally, and here the only difficult local case was that of a singular point of the ramification curve. Thus arose the problem of understanding the complement of the curve $f(x,y) = 0$ topologically in a small neighborhood of one of its singularities, and in particular of computing the fundamental group of this complement. If B_ε is the closed ε-neighborhood of $(0,0)$ in $\mathbb{C}^2$, then the set $\{f(x,y) = 0\} \cap B_\varepsilon$ is a cone over $\{f(x,y) = 0\} \cap S_\varepsilon$, where $S_\varepsilon = \partial B_\varepsilon$ is the 3-sphere, and this last is a link – a disjoint union of embedded circles. Designating these as algebraic links, the problem became to study and classify the algebraic links, and in particular to compute the fundamental groups of their complements. This problem also fit well into the then current development of knot theory.

Karl Brauner received the problem in this form from his thesis advisor Wirtinger, who, he says, had spoken of it in a seminar in Vienna as early as 1905. Brauner published what seems to be the first paper explicitly on algebraic links [B] in 1928. He describes these links, using stereographic projection to move them to R^3 from S^3, and gives an explicit presentation for the fundamental group.

The computational part of this, the stereographic projection, can easily be avoided, as was almost immediately pointed out by Kähler [K]; the idea is to replace the intersection with S_ε by the intersection with the "square sphere" $\{x,y \in C^2 \mid |x| = \varepsilon, |y| \leq \varepsilon \text{ or } |x| \leq \varepsilon, |y| = \varepsilon\}$. With a suitable coordinate choice one can arrange that $\{x,y \mid f(x,y) = 0\}$ meets this square sphere only in solid torus given by $|x| = \varepsilon$, $|y| \leq \varepsilon$, and it is then trivial to see that the Newton approximations given above describe successive cables on a torus knot.

It was shown by Werner Burau, first for knots [Bu 1] and then for algebraic links with ≤ 2 components [Bu 2] that the fundamental groups of the link components, and indeed even their Alexander polynomials, distinguish among them. (The same was proved for all algebraic links, by Evers, and Yamamoto, only recently [Ev], [Ya].) Burau's result was rediscovered by Zariski [Z], who computed the Alexander polynomial by considering the k-sheeted cyclic branched covers of the link complements, a procedure familiar from the algebraic geometry of the surfaces $z^k - f(x,y) = 0$.

It turns out that the data about the singularity of $f(x,y)$ at 0 which is contained in the topology of the associated link is fundamental to questions of algebraic geometry. It has now become essentially the basis of the theory of equisingularity developed over the last 20 years by Zariski [Z 2] and others. In the meantime other invariants of this topology have come to be studied. For example, Lê showed [Lê] that algebraic links are determined by their cobordism class.

One of the most important steps in the more recent study of algebraic links was the study of the fibration and monodromy, and the extension to higher dimensions, undertaken by Milnor [Mi 2]. The point is that if we regard the polynomial $f(x,y)$ as a function $f: \mathbb{C}^2 \to \mathbb{C}$, and restrict it to the preimage in B_ε of a very small disk D_δ about 0, then

$$f: (B_\varepsilon \cap f^{-1}D_\delta - f^{-1}(0)) \to D_\delta - \{0\}$$

is a fibration, and Milnor showed that the restriction of this fibration to the circle $S^1_\delta = \partial(D_\delta)$, which of course carries the same information, is isomorphic to a fibration of the exterior in S^3_ε of the algebraic link $\{f(x,y) = 0\} \cap S^3_\varepsilon$. The corresponding results are true in higher dimensions for complex polynomials with only an isolated singularity at the origin. Although there are no Puiseux or Newton expansions from which to extract information in the higher dimensional case, this fibration suggests that one can approach the topology of the link in any dimension by studying the fiber and the monodromy; this is the program of [Mi 2]. If F is the fiber of the fibration determined by a polynomial in $n+1$ variables, then Milnor showed that F has the homotopy-type of a bouquet of n-spheres, so that the most obvious discrete invariants are those of the Seifert form $S(-,-)$ on $H_n(F)$; given two n-cycles u and v on F, we push u off F in a positive normal direction to u_+, say, and set $S(u,v) = \ell(u_+,v)$, where ℓ denotes the linking number. In particular, the monodromy action on $H_n(F)$ and the intersection form on $H_n(F)$ can easily be deduced from S. Best of all, as A. Durfee observed in [Du], the theory developed by J. Levine in [Lev] to classify some higher dimensional knots actually shows that for all $n \geq 3$, the link associated to f is actually classified by its Seifert form.

Thus in the classical case of algebraic links in S^3, one has on one side a complete invariant, the characteristic pair, and on the other hand the Seifert form and its associated monodromy and intersection form, which become the "right" invariants in higher dimensions. For algebraic *knots* there were at least reasonable methods for constructing the monodromy, and thus the Seifert form from the characteristic pairs, [A'C], [Ga], but for links these methods did not produce useful information.

Our original motivation for the work reported here was to develop tools for computing the Seifert form to the extent possible, and also to understand whether algebraic links differed from general links, for which the Seifert form is often a very poor invariant, enough to give the Seifert form in this dimension the same significance for algebraic links that it has in higher dimension.

Several obvious test problems present themselves for such a program.

1) Which exteriors of iterated torus links fiber?

It has long been obvious that the algebraic links are a special case of the "iterated torus links." As we began our work, a substantial treatment by Sumners and Woods, of this more general class had just appeared [S-W]. The algebraic links all have, by Milnor's result, fibered exteriors. We wondered, as did Sumners and Woods, which iterated torus links have fibered exteriors (the answer mentioned at the end of that paper is unfortunately wrong), and to what extent those with fibered exteriors were "like" the algebraic ones.

2) Find the Jordan normal form of the monodromy from the characteristic pairs.

It was shown by Lê that the monodromy has finite order for algebraic knots [Lê 2], but A'Campo [A'C] and others had given examples of algebraic links with algebraic monodromy of infinite order. On the other hand, a theorem of Grothendieck [De] asserts that the monodromy coming from a polynomial in $n+1$ variables has Jordan blocks of size at most $n+1$, and the plane curve case $n = 1$ seemed a natural place in which to begin looking for more precise information.

3) Is the multiplicity of a polynomial $f(x_0, \cdots, x_n)$ with an isolated singularity at 0 (the degree of the lowest degree term of f) determined by the topological type of the link?

This question, attributed to Zariski, may, by the Durfee-Levine result, be interpreted as asking for a formula for the multiplicity in terms of the Seifert form as soon as $n \geq 3$. It is known that for $n = 1$, the answer to the question is yes. If one could give, in case $n = 1$, a formula for the multiplicity in terms of the Seifert form, this formula could perhaps be generalized, giving a conjecture in higher dimension of a much more definite sort. This problem leads us back, in any case to the obvious:

4) Calculate the Seifert form in terms of the characteristic pairs (or combinatorially equivalent information); does it determine the topological type of the link?

Our results are mixed. Problem 1) may be regarded as completely solved in this monograph; in fact the iterated torus links are shown to belong to a far more natural class of links, the "Graph links," and the problem is resolved in this more general setting. Problem 2) is also completely settled for algebraic links, though the obvious generalizations to iterated torus links and graph links are much less satisfactorily resolved. Problem 3) is not resolved here.

As for 4) our methods give a reasonably sharp analysis of the *real* Seifert form $R \otimes S(-,-)$, but, although they allow a systematic calculation of the integral Seifert form in concrete cases, they do not give a closed form expression for S in terms of the characteristic pairs, and we do not know how to give invariants characterizing S over Z. We also do not know whether the Seifert form determines the topological type of an algebraic link (it certainly does *not* determine the topological type of the more general graph links – even knots! – considered below).

As we remarked before, our method is to examine a certain canonical decomposition of the link exterior. In the case of algebraic links the exterior is put together out of Seifert-fibered pieces alone (this property is the definition of graph links). Though such geometric decompositions are generally and canonically available, as described below, we first understood them in the special case of algebraic links by examining the succession of Newton approximations given above. It seems worthwhile, before launching into the general theory, to exhibit this decomposition in a sample special case directly, as we initially understood it.

Consider then, the case of an iterated torus knot corresponding to the Newton approximations

$$y_1 = x^{3/2}$$

$$y = y_2 = x^{3/2}(1 + x^{1/4}),$$

which is the "solution" to the polynomial equation

$$f(x,y) = y^4 - 2x^3y^2 - 4x^5y + x^6 - x^7 = 0$$

(see Appendix to Chapter I).

Taking $|x| = \varepsilon$ with ε small, we see that y_1 traverses a $(2,3)$ torus knot K_1 inside the solid torus $|y| \leq \varepsilon$, while y_2 runs along in a perturbation K_2 of this torus knot, easily seen to be the $(2,13)$ cable knot on K_1 (see Prop. 1A.1 below). If ε is sufficiently small, then K_2 will lie in a tubular neighborhood N of K_1.

The link exterior is now easy to describe. It consists of two pieces: the exterior of K_1 joined along a torus, its boundary, to a solid torus from which a tubular neighborhood of the $(2,13)$ knot has been removed. Note that a solid torus is the exterior in S^3 of the unknot, so this second piece is actually the exterior of a two-component link whose components are an unknot and a $(2,13)$ knot, linking twice:

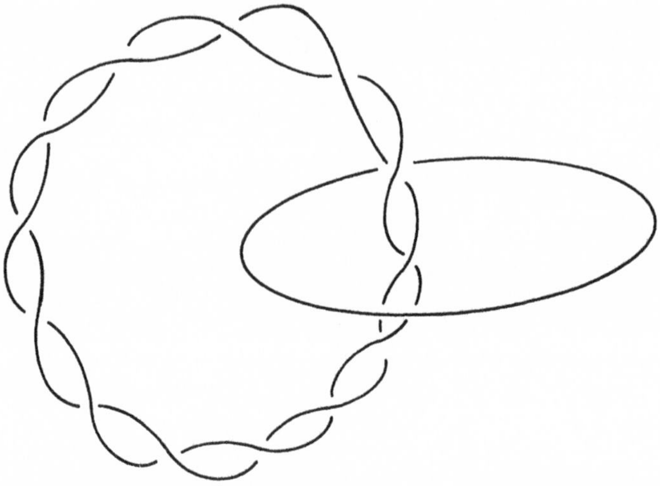

The torus along which this piece is glued to the first piece may be described as the boundary of a tubular neighborhood of the un-knot component. It turns out that this sort of decomposition is canonically available for any link.

In the example just given, as for any algebraic link, both pieces in the decomposition are Seifert fibered:

For the first piece, the restriction to S^3 of the map

$$\mathbb{C}^2 \to \mathbb{P}^1_{\mathbb{C}} = S^2 : (z,w) \to z^2/w^3$$

is a Seifert fibration of S^3 with exceptional fibers a pair of circles and with general fiber a $(2,3)$ torus knot

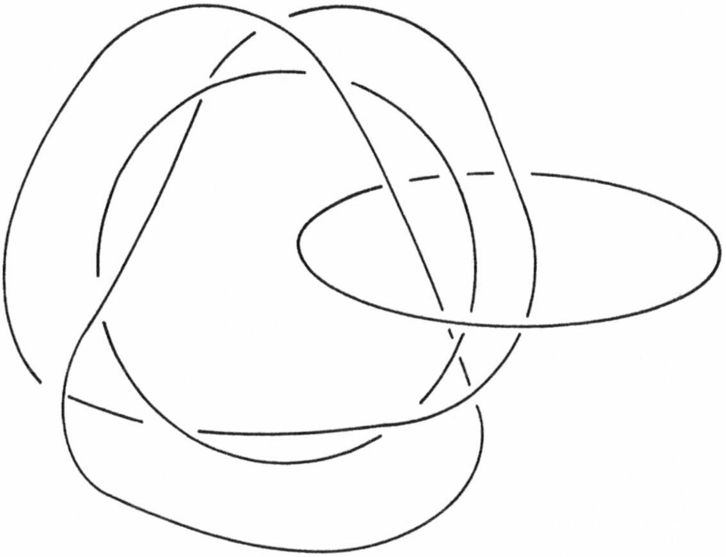

The first piece is obviously S^3 minus the preimage under this map of S^2 minus a disk in S^2 not meeting 0 or ∞.

Similarly, for the second piece, the restriction to the solid torus $|z| \leq 1$ of the map

$$\mathbb{C}^2 \to \mathbb{P}^1_{\mathbb{C}} = S^2 : (z,w) \to z^2/w^{13}$$

makes our second piece a Seifert bundle over the twice punctured sphere, with one exceptional fiber.

Perhaps the most interesting thing about this decomposition of the exterior of K_2 is that the natural Milnor fibration $S^3_\varepsilon - K_2 \to S^1 : (x,y) \mapsto f(x,y)/|f(x,y)|$, which fibers $S^3 - K_2$ by surfaces over S^1, restricts to a fibration of each of the pieces – that is, every fiber is transverse to the torus separating the two pieces.

In fact, it is easy to give the fibration on each piece separately, and then paste the two together to give the fibration of the exterior of K_2. Such a result, for all algebraic knots, is the essential point of the paper [A'C] of A'Campo. But as we show below, it is quite general: fibrations of link exteriors always come from fibrations of the individual pieces. This is one aspect of the kind of additivity which holds for the decomposition of link exteriors treated here.

Preview

In the course of our study of algebraic links – that is, links of singularities of algebraic plane curves – we became convinced that a central role should be played by a certain canonical decomposition of the link exterior into pieces derived from the exteriors of simple torus knots. This decomposition seemed to us at first highly dependent on the way in which algebraic links are built up inductively by repeated cabling operations, and our first preprint on the subject [“Toral links”] contained just such an inductive treatment of the decomposition. We subsequently realized that an analogous *splice decomposition* of the link exterior is available for every link from the Jaco-Shalen-Johannson decomposition of 3-manifolds. In the general case, each piece of the decomposition is either Seifert-fibered or simple. The special features of the algebraic case which allow an excellent grip on the invariants of algebraic links are mostly available as soon as the pieces in the splice decomposition are all Seifert-fibered.

In Chapter I of this monograph we exploit the Jaco-Shalen-Johannson theory, as sketched above, to give a foundation for the theory of links (in homology 3-spheres) based on splice decompositions. In its current form, our presentation of the splice decomposition is adapted from that used by Siebenmann in his study of homology 3-spheres [Si]; the case he treats corresponds in our setting to the case of an empty link. The key point is perhaps that a large number of standard invariants (genus, Alexander polynomial, fiberability, ...) are in a suitable sense additive under splice decomposition. Since splice decomposition generalizes many standard constructions of links, such as forming cables, connected sums, twisted doubles, etc., many of the computations in the literature for the behavior of the Alexander polynomial, Seifert form, fiberability, etc., are subsumed.

The splice components of a link are, in a natural way *multilinks* – that is, links for which each link component is assigned an integer weight (this has the familiar interpretation in the case of algebraic links). We thus work mostly with multilinks.

In Chapter II-V, we analyze the case of links – which we call graph links – for which all the splice components have Seifert-fibered structures ("Seifert links") as is the case with links of plane-curve singularities. In terms of the associated "splice diagram," which exhibits all the topological data of the decomposition in this case, we are able to give a complete classification of these links (characterizing the algebraic ones among them), and by exploiting the additivity mentioned above, we are able to give simple computations of a number of important algebraic invariants. In the case of links of algebraic plane-curve singularities, the splice diagram is closely related to the set of Puiseux expansions of the branches of the curve, so these computations can be interpreted as computations from those Puiseux series. We obtain, for example, a formula for the Jordan normal form of the algebraic monodromy, and thus a simple criterion for the finiteness of that monodromy, in these terms.

In Chapter V we show how the results on graph links can be translated into the language of plumbed manifolds and give effective algorithms for converting between plumbing graphs and splice diagrams; in the algebraic case, this corresponds to the translation between "resolution graph" and Puiseux series. An efficient way of computing the characteristic pairs of a given polynomial is to resolve its singularities and apply the technique of Chapter V to the resolution graph.

We now describe splicing, and the contents of this monograph, in more detail.

A *link* L is for us a pair (Σ,K) where Σ is an oriented (PL) homology 3-sphere and K is a disjoint union of oriented circles in Σ, called the link components. Given two links $L=(\Sigma,K)$ and $L'=(\Sigma',K')$, and a component $S \subset K$ and $S' \subset K'$ of each we construct the splice

$$L \underset{S}{\quad} \overline{\qquad\qquad} \underset{S'}{\quad} L'$$

of L and L′ by pasting together complements of tubular neighborhoods of S and S′, matching meridian to longitude and longitude to meridian. The link components of L and L′ other than S and S′ are the link components in the result.

This construction, together with its special cases cabling and connected sum, are treated in section 1. The Appendix to Chapter I contains a review of the algebraic case, and gives the construction of the splice diagram of an algebraic link from the Puiseux series (splice diagrams in general are not taken up until section 8). The reader interested solely in the algebraic case can skip to section 9 after reading section 1 and this Appendix.

Section 2 contains a treatment of desplicing – that is, writing a given link as a splice – based on the Jaco-Shalen-Johannson theory. In particular, it is shown that every link admits an essentially unique splice-decomposition into simple and Seifert-fibered pieces.

In sections 3 to 5 we describe how to understand the topology and invariants of a link in terms of its splice components. We do this for the genus of a link (or rather its appropriate substitute for multilinks: Thurston's norm on homology, Theorem 3.3), fiberability (Theorem 4.2), the geometric monodromy in the fibered case (Addendum to 4.2), the characteristic polynomial of the monodromy (Theorem 4.3), and the single variable and multivariable Alexander polynomials (Theorems 5.2 and 5.3). These results can be summarized by saying that with appropriate normalizations, these invariants are essentially additive over the splice components. Many of the results of these sections apply more generally to the Jaco-Shalen-Johannsen splitting of any 3-manifold, rather than just a link complement, and we describe this too. Section 4 also contains a discussion of the uniqueness of the fibration, and of minimal Seifert surfaces, of a fibered link. Section 6 discusses some necessary algebra of fitting ideals associated to abelian covers and is of interest in its own right.

Chapter II (sections 7 to 9) describes the classification first of Seifert links, then of general graph links, then the special case of solvable links. We classify graph links by certain "splice diagrams," which code the splice components and how they are spliced together. Such diagrams first appeared in Seibenmann [Si] to classify plumbed homology spheres (empty graph links!). We discovered them independently and roughly simultaneously, as a way to classify graph links by coded linking information. We were using a plumbing approach (see Chapter V) which was less transparent; the idea of using splicing was adapted from Siebenmann.

In Chapter III (sections 10-15) we compute all the earlier mentioned invariants for graph links, most importantly the Alexander polynomial. Moreover, in the fibered case we also compute the Jordan normal form of the algebraic monodromy. The eigenvalues are roots of unity and only 1×1 and 2×2 Jordan blocks occur (for plane curve singularities this is a result of Grothendieck). Surprisingly the eigenvalue 1 only has 1×1 Jordan blocks; a feature which is not generally valid either for more general links or for complex hypersurface singularities in higher dimensions.

Even in the previously known cases (toral links were done in [E-N] and the Alexander polynomial was treated for a class of toral links including algebraic links in Sumners and Woods [S-W]), our computations differ from earlier work in that they give simple closed formulae in terms of data of clear geometric meaning (certain linking numbers), rather than inductive algorithms as in [S-W] and [E-N].

The computation of the Jordan normal form for the algebraic monodromy is based on a complete geometric description of the monodromy in sections 11 and 13. This description makes it routine (but tedious) to write down an explicit monodromy matrix in any particular example, but given the results one gets, it does not seem worth the effort; there is little one can see by staring at a large integer matrix. More usefully, we can compute enough signature invariants to determine the equivalence class of the monodromy as an isometry of the intersection form (this is equivalent to knowing the Seifert form over $\mathbf{R}$); this is done from a plumbing approach in [Ne 4] and via splicing in [Ne 5]. Ultimately one would wish to find and compute enough invariants to classify such data over $\mathbf{Z}$.

We can explicitly write the Seifert form of the result of splicing a knot to a link in terms of the pieces. In particular this gives the Seifert form for any knot in terms of its splice components, but we do not have a closed formula for arbitrary links. We do this in section 15, and also give examples which stress that, even among solvable knots, many distinct knots can have the same Seifert form.

Chapter IV is on a lighter note. In section 16 we give a classification which is included mainly for its amusement value: fibered links in S^3 with trivial monodromy. Here is a picture of the fiber of such a link; a different picture of the same fiber is in section 16.

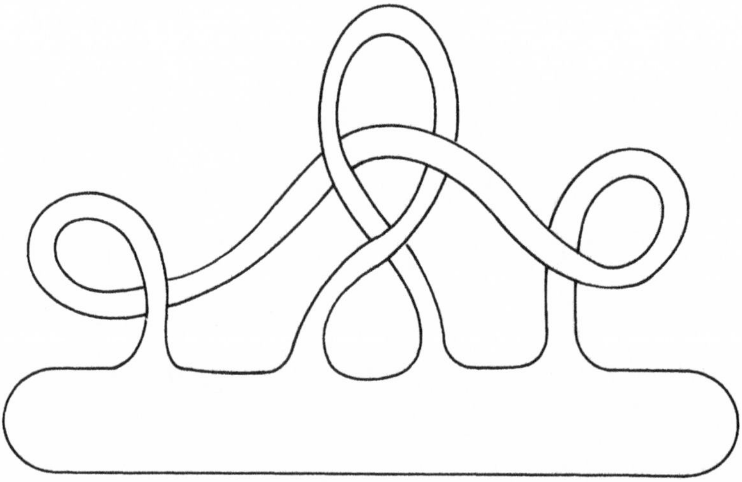

Section 17 contains a collection of other examples, with their splice diagrams; in particular, the reader will find there a large collection of fibered links with trivial algebraic monodromy, but non-trivial geometric monodromy.

The final chapter, Chapter V, describes the plumbing approach to graph links. Plumbing originally arose through resolution of singularities and in section 19 we review the resolution of a plane curve singularity and how this leads to a plumbing description of its link. We give algorithms to get a splice diagram from a plumbing description (section 20) and vice versa (section 22). These algorithms are computationally quite efficient and are based on a simple continued fraction algorithm to diagonalize the matrix of a graph, described in section 21. The basic invariants of a graph link are given in terms of plumbing in section 22. This extends previous work of A'Campo and M. Evers. The final section describes how to recognize algebraicity of a link in terms of either plumbing or its splice diagram.

Chapter I

FOUNDATIONS

1. *Basic constructions*

In this section we describe various natural operations such as connected sum, disjoint sum, cabling, that can be used to construct new links from old. As we shall see, they are all special cases of one operation, which we call splicing. However the operation of splicing often results in links in homology spheres rather than links in the 3-sphere S^3. It is therefore most convenient to permit homology spheres as the ambient spaces for links from the start.

By a *link* $L = (\Sigma, K) = (\Sigma, S_1 \cup \cdots \cup S_m)$ we shall mean a pair consisting of an oriented homology 3-sphere Σ and a collection of smooth disjoint oriented simple closed curves S_i in Σ. If the ambient homology sphere Σ is the standard sphere S^3, we speak of a *link in* S^3. If the S_i have not been oriented, we speak of an *unoriented link.*

There are two natural concepts of "sum" of links. If $L = (\Sigma, S_1 \cup \cdots \cup S_m)$ and $L' = (\Sigma', S_1' \cup \cdots \cup S_n')$, we can form the *disjoint sum*

$$L + L' = (\Sigma \# \Sigma', S_1 \cup \cdots \cup S_m \cup S_1' \cup \cdots \cup S_n')$$

by taking connected sum of Σ and Σ' along disks which do not intersect the links. We can also form a *connected sum*

$$L \# L' = (\Sigma \# \Sigma', (S_1 \# S_1') \cup S_2 \cup \cdots \cup S_m \cup S_2' \cup \cdots \cup S_n')$$

by taking connected sum of Σ and Σ' along disks which intersect the links in intervals $I \subset S_1$ and $I' \subset S_1'$. We write

$$L \# L' \text{ (along } S_1, S_1')$$

to emphasize which link components we are summing along if necessary.

If the link L can be expressed as a nontrivial disjoint sum (that is, neither summand is the empty link in S^3), we say L is a *reducible* link, otherwise L is *irreducible*. Thus $L = (\Sigma, K)$ is reducible if and only if the link complement $\Sigma - K$ is a reducible 3-manifold. A link in S^3 is reducible if and only if it is a split link. Since, by Milnor [Mi 1], any link is the disjoint sum of irreducible links other than (S^3, ϕ) in an essentially unique way, for most purposes no generality is lost by considering only irreducible links.

If $L = (\Sigma, K) = (\Sigma, S_1 \cup \cdots \cup S_m)$ is a link, $N(K) = N(S_1) \cup \cdots \cup N(S_m)$ shall denote a closed tubular neighborhood of K in Σ. $\Sigma - \mathrm{int}(N(K))$ will be called the *link exterior*, as opposed to the *link complement* $\Sigma - K$, though in most situations this distinction will be academic.

We shall denote by M_i and L_i a "topologically standard" meridian and longitude of the link component S_i. They are the pair of oriented simple closed curves in $\partial N(S_i)$ which are determined up to isotopy by the homology and linking relations:

$$M_i \sim 0\,,\ L_i \sim S_i \ \text{ in } H_1(N(S_i))\,;$$

$$\ell(M_i, S_i) = 1\,,\ \ell(L_i, S_i) = 0\,.$$

Here $\ell(-,-)$ denotes linking number in Σ.

If p and q are relatively prime integers, we denote by $S_i(p,q)$ the unique (up to isotopy) oriented simple closed curve in $\partial N(S_i)$ homologous to $pL_i + qM_i$. More generally $dS_i(p,q) = S_i(p,q)_1 \cup \cdots \cup S_i(p,q)_d$ denotes the disjoint union of d parallel copies of $S_i(p,q)$ in $\partial N(S_i)$. The operation of replacing $L = (\Sigma, K)$ by either $(\Sigma, K \cup dS_i(p,q))$ or $(\Sigma, K \cup dS_i(p,q) - S_i)$ is called a *cabling operation*. As already mentioned, the above operations are all special cases of one operation, splicing, which we now describe. Let $L = (\Sigma, K)$ and $L' = (\Sigma', K')$ be links and choose components $S \subset K$ and $S' \subset K'$. Let $N(S)$ and $N(S')$ be tubular neighborhoods and $M, L \subset \partial N(S)$ and $M', L' \subset \partial N(S')$ be standard meridians and longitudes. Form

$$\Sigma'' = (\Sigma - \text{int } N(S)) \cup (\Sigma' - \text{int } N(S')) ,$$

pasting along boundaries by matching M to L' and L to M'. An easy Mayer Vietoris sequence argument shows Σ'' is again a homology sphere. Since this pasting map $\partial N(S) \to \partial N(S')$ reverses orientation, Σ'' inherits the same orientation from each of Σ and Σ'.

DEFINITION. The link $(\Sigma'', (K-S) \cup (K'-S'))$ is called the *splice of* L *and* L' *along* S *and* S', denoted

$$L \text{———} L' ,$$

or

$$\underset{S\quad S'}{L \text{———} L'}$$

depending on whether the components along which the splicing is being done are clear from context or not.

To see how the various operations on links are related, first note that when we splice links L and L' along components S and S', the standard meridian and longitude of any other component of L or L' remain unchanged. Hence splicing commutes with any of our basic operations (disjoint or connected sum, cabling, splicing) in the sense that if we first perform a splice and then perform another operation we get the same result as by doing the two operations in the reverse order.

Note also that splicing any link $L = (\Sigma, K)$, along $S \subset K$, to the following link in S^3,

along S_1, just gives us L back again with S replaced by S_2. Indeed, that this is true up to orientation is clear, since we can choose $N(S_1)$ and $N(S_2)$ so that $S^3 - \text{int}\, N(S_1) = N(S_2)$. Orientation is also easily checked (for instance by Proposition 1.2 below).

These remarks make the following proposition clear.

PROPOSITION 1.1.

1) *Let* $(S^3, S_1 \cup S_2)$ *be as above. Let* S *be a component of a link* $L = (\Sigma, K)$. *Cabling on* S *can be described via splicing by*

$$(\Sigma, K \cup d\, S(p,q)) \cong L \underset{S \quad S_1}{\text{———}} (S^3, S_1 \cup S_2 \cup d\, S_2(p,q))$$

$$(\Sigma, K \cup d\, S(p,q) - S) \cong L \underset{S \quad S_1}{\text{———}} (S^3, S_1 \cup d\, S_2(p,q))$$

2) *Let* $L^{(0)} = (S^3, S_0 \cup \cdots \cup S_n)$ *be the following link*

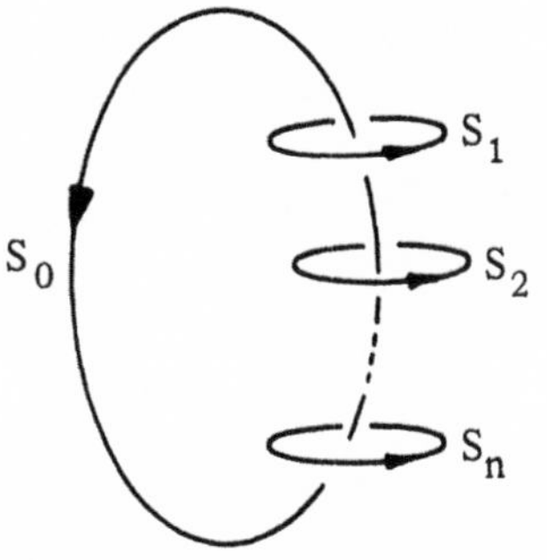

Given links $L^{(i)} = (\Sigma^{(i)}, K^{(i)})$ *and components* $S^{(i)} \subset K^{(i)}$ *for* $i = 1, \cdots, n$, *the connected sum of the* $L^{(i)}$ *along the* $S^{(i)}$ *is*

$$\mathop{\#}_{i=1}^{n} L^{(i)} \text{ (along } S^{(i)}) \cong L^{(0)}$$

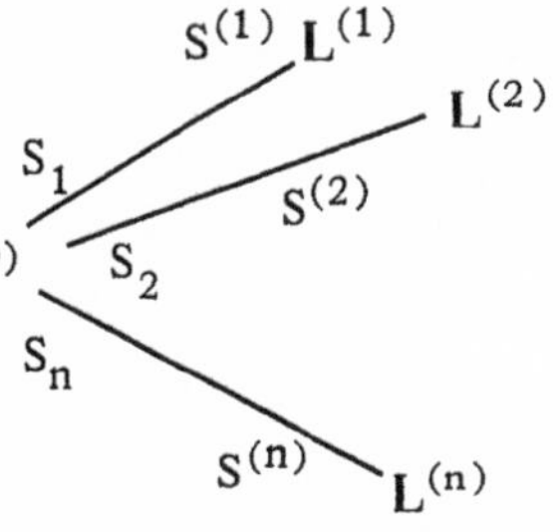

3) *If we delete* S_0 *in part 2) we get a splice description of disjoint sum.*

Indeed, 1) is clear from our previous remarks and 3) follows from 2). To see 2), observe that $\mathbb{L}^{(0)}$ is the connected sum of the following links along $S_0^{(1)}, \cdots, S_0^{(n)}$:

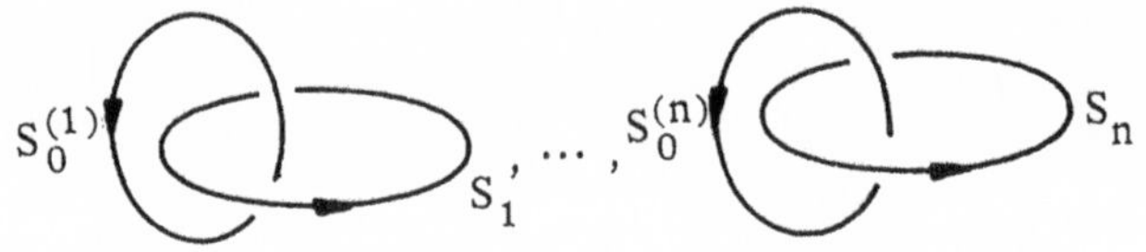

The following simple proposition on linking numbers will be very useful later.

PROPOSITION 1.2. *If*

$$(\Sigma'', K'') = (\Sigma, K) \underset{S \quad S'}{\text{———}} (\Sigma', K')$$

and S_1 *and* S_1' *are components of* $K-S$ *and* $K'-S'$ *then*

$$\ell_{\Sigma''}(S_1, S_1') = \ell_\Sigma(S_1, S) \cdot \ell_{\Sigma'}(S', S_1'),$$

where $\ell_\Sigma, \ell_{\Sigma'}, \ell_{\Sigma''}$ *mean linking number in* Σ, Σ', Σ''.

Proof. Put $p = \ell_\Sigma(S_1, S)$ and $q = \ell_{\Sigma'}(S', S_1')$. By definition of linking numbers, the standard longitude L_1 of S_1 is homologous in $\Sigma - \mathrm{int}(N(S_1) \cup N(S))$ to p times the meridian M of S. Similarly we have the homology relation $L' = qM_1'$ in $\Sigma' - \mathrm{int}(N(S') \cup N(S_1'))$. Thus in $\Sigma'' - \mathrm{int}(N(S_1) \cup N(S_1'))$ we have the homology relation $L_1 = pM = pL' = pqM_1'$, so $\ell_{\Sigma''}(S_1, S_1') = pq$, as claimed.

2. *Splice decomposition*

Define a *Seifert link* to be a link $\mathbb{L} = (\Sigma, K)$ whose exterior Σ–int $N(K)$ admits a Seifert fibration. As usual, a *simple link* is defined to be an irreducible link $L = (\Sigma, K)$ with the property that any incompressible torus in Σ–int $N(K)$ is boundary parallel. By Thurston's hyperbolization theorem [Th 1] this is equivalent to saying that the link complement $\Sigma - K$ admits a complete hyperbolic structure of finite volume, except possibly if $K = \phi$ and Σ is not sufficiently large.

The splitting theorem of Jaco and Shalen [J-S] and Johannson [Jo], as formulated in [J-S], implies that any irreducible link $L = (\Sigma, K)$ can be expressed as the result of splicing together a collection of Seifert links and simple links, and moreover the minimal way of doing this is "essentially" unique. We must explain what we mean by this.

PROPOSITION 2.1. *If* $L'' = (\Sigma'', K'')$ *is a link and* $T^2 \subset \Sigma'' - K''$ *is an embedded torus, then* L'' *is the result of a splicing operation along this* T^2, $(\Sigma'', K'') = (\Sigma, K) \underset{S \quad S'}{\overline{\qquad\qquad}} (\Sigma', K')$, *which is uniquely determined up to reversing the orientations of both* S *and* S'. *Moreover, if* $\Sigma'' \cong S^3$, *then* $\Sigma \cong S^3$ *and* $\Sigma' \cong S^3$.

Proof. By Alexander duality, T^2 cuts Σ'' into two pieces, call them Σ_0 and Σ_0', which are both homology circles. Moreover there is a unique basis $\{M, M'\}$ of $H_1(T^2; Z)$, up to changes of sign, such that M bounds in Σ_0' and M' bounds in Σ_0. Choose signs so that $\{M', M\}$ orients T^2 as $\partial \Sigma_0$. Let $\Sigma = \Sigma_0 \cup (D^2 \times S^1)$, pasted along boundaries so that M is nullhomologous in $D^2 \times S^1$, and let S be the core circle in $D^2 \times S^1$, oriented parallel to M'. Similarly construct $\Sigma' = \Sigma_0' \cup (D^2 \times S^1)$ and S' by exchanging the roles of M and M'. Then $L = (\Sigma, (K'' \cap \Sigma_0) \cup S)$ and $L' = (\Sigma', (K'' \cap \Sigma_0') \cup S')$ are the desired links, whose splice is L''.

To see the truth of the final sentence of the theorem, observe that if $\Sigma_0 \cong D^2 \times S^1$, then $\Sigma' \cong \Sigma''$ and $\Sigma \cong S^3$. But if $\Sigma'' \cong S^3$ then either Σ_0 or Σ_0' is diffeomorphic to $D^2 \times S^1$ by Dehn's lemma [Pa].

The main result of this section is now just an interpretation of the Jaco-Shalen Johannson splitting theorem for our situation:

THEOREM 2.2. *Let* $L = (\Sigma, K)$ *be an irreducible link. Then there exists a collection* $\{T_i\}$ *of disjoint incompressible non-boundary-parallel embedded tori in the link exterior* $\Sigma - \mathrm{int}\ N(K)$ *such that if one "de-splices"* L, *as in Proposition 2.1, along all these tori, each resulting link is either a Seifert link or a simple link. Moreover, a minimal such collection of tori is included up to isotopy in any such collection, and is therefore unique.*

Indeed, the quoted splitting theorem guarantees a minimal splitting, unique in the sense of the above theorem, of any irreducible compact oriented three-manifold $(M, \partial M)$, along incompressible non-boundary-parallel tori and annuli, into "Seifert pairs" and "simple pairs." Here "Seifert" means either Seifert fibered in the usual sense, or homeomorphic to the total space E of an I-bundle over a surface, with $E \cap \partial M$ equal to the total space of the ∂I-bundle. Annuli can occur in the splitting surface only if some component of ∂M has genus at least 2. Thus I-bundle pieces can occur only in this case also. Thus in our situation the splitting is only along tori and "Seifert fibered" has its usual meaning, so the theorem follows.

3. *Multilinks and genus*

In this section we discuss the genus of a link in terms of the splice decomposition. As we point out, much of what we do applies more generally to arbitrary 3-manifolds. The concept of the genus of a knot does not generalize well to links, so we use an equivalent concept, Thurston's pseudonorm on homology, which we explain later.

For the discussion of this section and the following sections and for many other purposes too (abelian invariants of links, links arising in algebraic geometry, etc.), it is natural to generalize the concept of link to a suitable concept of "link with multiplicities", which we call a multilink.

Let $(\Sigma, \mathcal{K})$ be an *unoriented* link and let $L = (\Sigma, S_1 \cup \cdots \cup S_n)$ be a link obtained by orienting $(\Sigma, \mathcal{K})$. By a *multilink on* $(\Sigma, \mathcal{K})$ we mean L together with an integer multiplicity m_i associated with each component S_i, with the convention that a component S_i with multiplicity m_i means the same think as $-S_i$ (S_i with reversed orientation) with multiplicity $-m_i$. We write the multilink

$$L(m_1, \cdots, m_n) = (\Sigma, m_1 S_1 \cup \cdots \cup m_n S_n) .$$

A link is thus simply a multilink with all multiplicities ± 1.

A multilink on $(\Sigma, \mathcal{K})$ as above determines an integral cohomology class $\underline{m} \in H^1(\Sigma - \mathcal{K}) = H_1(\Sigma - \mathcal{K})^*$ as follows: $\underline{m}$ evaluated on a 1-cycle S is the linking number

$$\underline{m}(S) = \ell(m_1 S_1 + \cdots + m_n S_n, S) = \sum_{i=1}^{n} m_i \ell(S_i, S) .$$

By Alexander duality the n linear forms $\ell(S_i, -) \in H_1(\Sigma - \mathcal{K})^*$ are a basis of $H_1(\Sigma - \mathcal{K})^* = H^1(\Sigma - \mathcal{K})$. Thus if we use this basis to identify $H^1(\Sigma - \mathcal{K})$ with Z^n, then $\underline{m} = (m_1, \cdots, m_n)$. We may thus also write $L(\underline{m})$ for $L(m_1, \cdots, m_n)$. The notation $L(\underline{m})$ has the advantage that in it, L can symbolize the unoriented link $(\Sigma, \mathcal{K})$, while the notation $L(m_1, \cdots, m_n)$ needs the identification $H^1(\Sigma - \mathcal{K}) = Z^n$, that is an orientation of $\mathcal{K}$. Given $\underline{m}$ and L one computes the m_i by $m_i = \underline{m}(M_i)$, M_i a standard oriented meridian of S_i.

If $N(\mathcal{K})$ is a tubular neighborhood of $\mathcal{K}$ in Σ, so $\Sigma_0 = \Sigma - \operatorname{int} N(\mathcal{K})$ is the link exterior, we identify $H^1(\Sigma - \mathcal{K}) = H^1(\Sigma_0)$ and consider $\underline{m}$ also as an element of $H^1(\Sigma_0)$.

Now suppose the unoriented link $L = (\Sigma, \mathcal{K})$ is the result of splicing unoriented links $L' = (\Sigma', \mathcal{K}')$ and $L'' = (\Sigma'', \mathcal{K}'')$, say

$$(\Sigma, \mathcal{K}) = (\Sigma', \mathcal{K}') \underset{S' \quad S''}{\overline{\qquad\qquad}} (\Sigma'', \mathcal{K}'') ,$$

where S', S'' are oriented components of $\mathcal{K}'$, $\mathcal{K}''$ respectively. Let

Σ_0, Σ'_0, Σ''_0 be the respective link exteriors, so $\Sigma_0 = \Sigma'_0 \underset{T^2}{\cup} \Sigma''_0$ by a suitable pasting. Then any cohomology class $\underline{m} \in H^1(\Sigma_0)$ restricts to classes $\underline{m}' \in H^1(\Sigma'_0)$ and $\underline{m}'' \in H^1(\Sigma''_0)$. We say $L(\underline{m})$ *is the result of splicing* $L'(\underline{m}')$ and $L''(\underline{m}'')$, written

$$L(\underline{m}) = L'(\underline{m}') \text{———} L''(\underline{m}''),$$

and we call $L'(\underline{m}')$ and $L''(\underline{m}'')$ splice summands of $L(\underline{m})$. The splice summands which result from the natural splice decomposition of section 2 are called the *splice components of* $L(\underline{m})$.

REMARK. Given components S', S'' of multilinks $L'(\underline{m}')$ and $L''(m'')$, the splicing

$$L'(\underline{m}') \underset{S' \quad S''}{\text{————}} L''(\underline{m}'')$$

is not usually defined. Indeed, in the above notation, the condition is that $\underline{m}'$ and $\underline{m}''$ must be restrictions of some cohomology class $\underline{m}$ on the link exterior Σ_0 for L. That is, $\underline{m}'$ and $\underline{m}''$ must restrict to the same class on the torus T^2 along which splicing is done. Thus, letting L' and L'' denote standard longitudes of S', S'' in L', L'', the condition of splice-ability is precisely

$$\underline{m}'(L') = m'' \text{ and } \underline{m}''(L'') = m',$$

where m', m'' are the multiplicities of S', S'', in $L'(\underline{m}')$, $L''(\underline{m}'')$.

By a *Seifert surface* for the multilink $L(\underline{m}) = (\Sigma, m_1 S_1 \cup \cdots \cup m_n S_n)$ we mean the natural generalization of the usual concept of Seifert surface. That is, it is an embedded oriented surface $F_0 \subset \Sigma - \mathcal{K}$ such that $\overline{F}_0 - F_0 \subset \mathcal{K}$ and $\overline{F}_0$ intersects a tubular neighborhood $N(S_i)$ of the component S_i as follows for each i :

$m_i \neq 0$

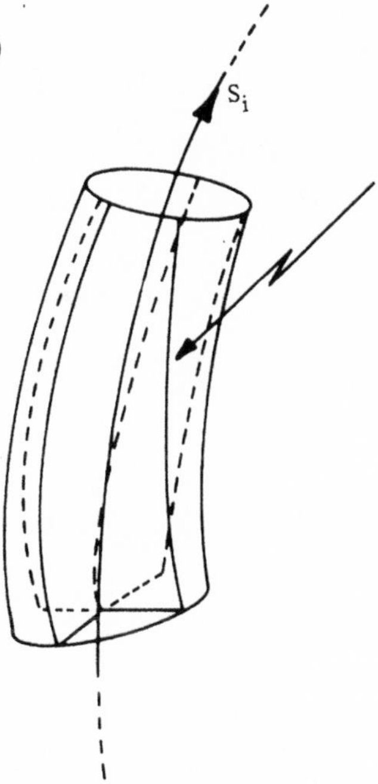

$F_0 \cap N(S_i)$ consists of $|m_i|$ leaves meeting along S_i. Orientation of F_0 is such that $\partial F_0 \cap N(S_i) = m_i S_i$ (thus there is no homological cancellation in ∂F_0 along S_i).

$m_i = 0$

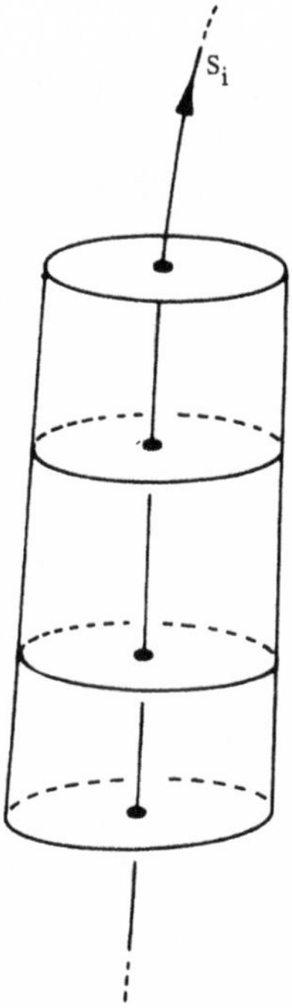

$\overline{F}_0 \cap N(S_i)$ consists of discs transverse to S_i. Orientation of F_0 is such that intersection number of S_i with each of these discs is the same (namely either $+1$ for each disc or -1 for each disc).

Let $\Sigma_0 = \Sigma - \mathrm{Int}\, N(\mathcal{K})$ be the link exterior. We also call $F = F_0 \cap \Sigma_0$ a Seifert surface, since it determines and is determined by F_0 up to isotopy. It has the following characterizing properties:

(i) F *is an oriented surface, properly embedded in* Σ_0 *(that is* $F \cap \partial\Sigma_0 = \partial F$ *transversally).*

(ii) $F \cap \partial N(S_i) = d_i S_i(p_i, q_i)$, *a* $d_i(p_i,q_i)$*-cable on* S_i, *where* d_i, p_i, q_i *are determined by:* $\gcd(p_i,q_i) = 1$, $d_i p_i = m_i$, $d_i q_i = -\ell(\sum_{j \neq i} m_j S_j, S_i)$.

The only nonobvious part of this is the formula for $d_i q_i$, which follows from the fact that F is a homology between $\sum_{j\neq i} m_j S_j$ and $-m_i L_i - d_i q_i M_i$, where L_i and M_i are standard longitude and meridian of S_i.

LEMMA 3.1. *The homology class* $[F] \in H_2(\Sigma_0, \partial\Sigma_0)$ *is dual to* $\underline{m} \in H^1(\Sigma_0)$. *Conversely, if* $F \subset \Sigma_0$ *is a properly embedded surface dual to* $\underline{m}$, *then, after suitably modifying* F *in a small neighborhood of* $\partial\Sigma_0$ *to "minimize"* $F \cap \partial\Sigma_0$, F *becomes a Seifert surface for* $L(\underline{m})$.

Proof. If F is a Seifert surface, the statement that F is dual to $\underline{m}$ is the statement: $F.S = \underline{m}(S)$ for any 1-cycle S in Σ_0, where $F.S$ is intersection number. But $F.M_i = m_i = \underline{m}(M_i)$ for each i, and the M_i form a basis of $H_1(\Sigma_0)$.

For the converse we shall need the following well-known lemma, whose proof we omit.

LEMMA 3.2. *If* C *is a collection of disjoint oriented simple closed curves in* $T^2 = S^1 \times S^1$, *then for some coprime* $p, q \in Z$,

$$C = C_1 \cup \cdots \cup C_k \cup -C_{k+1} \cup \cdots \cup -C_{k+\ell} \cup C_1' \cup \cdots \cup C_m'$$

with C_i *isotopic to* $S(p,q) = \{(t^p, t^q) \mid t \in S^1\}$ *for* $i = 1, \cdots, k+\ell$, *and* C_j' *a curve which bounds a disk in* $T^2 - (C_1 \cup \cdots \cup C_{k+\ell} \cup C_1' \cup \cdots \cup C_{j-1}')$ *for* $j = 1, \cdots, m$.

Returning to Lemma 3.1, suppose $F \subset \Sigma_0$ is properly embedded and dual to $\underline{m}$. The homology class of $F \cap \partial N(S_i)$ in $\partial N(S_i)$ is dual to the restriction of $\underline{m}$ to $\partial N(S_i)$. We have already computed this homology class, it is $F \cap \partial N(S_i) \sim d_i(p_iL_i + q_iM_i)$ in our earlier notation. It suffices to show we can alter F to make $F \cap \partial N(S_i)$ isotopic in $\partial N(S_i)$ to $d_iS_i(p_i,q_i)$ for each i.

Represent $F \cap \partial N(S_i)$ as in Lemma 3.2. Now, working inside a collar neighborhood of $\partial N(S_i)$ in Σ_0, eliminate the C_j' iteratively for $j = m,\ m-1, \cdots, 1$ by displacing C_j' into the interior of Σ_0 and then spanning in a disk. Then cancel neighboring pairs of oppositely oriented C_j's iteratively by displacing into the interior of Σ_0 and spanning in annuli. When no further such cancellation is possible, $F \cap \partial N(S_i)$ will be an isotopic copy of $d_iS_i(p_i,q_i)$, which can be straightened out, if desired, by an isotopy of F within a collar neighborhood of $\partial N(S_i)$:

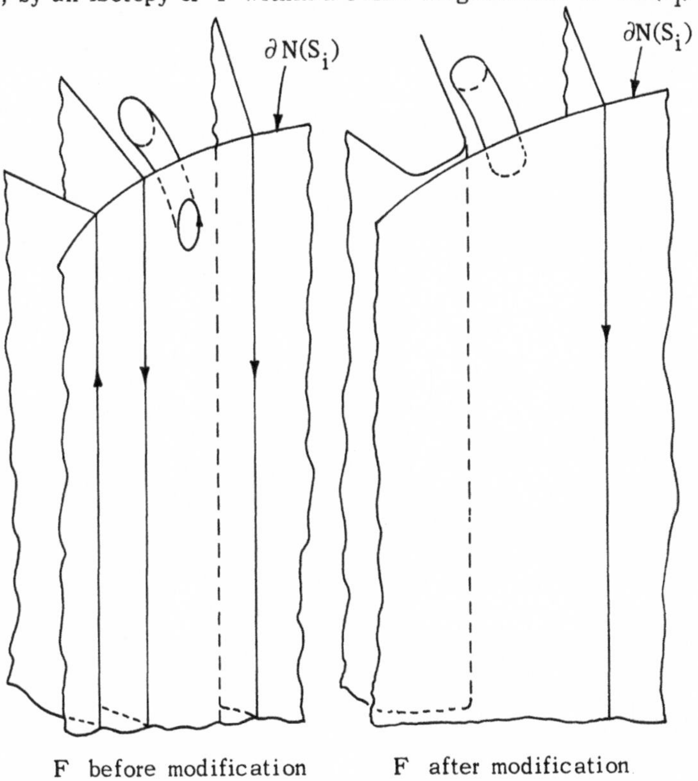

F before modification F after modification

Following Thurston [Th 2] we define, for a connected compact surface F,

$$\chi_-(F) = \max(0, -\chi(F)) \, ,$$

and more generally, for any compact surface F, define $\chi_-(F)$ as $\Sigma\chi_-(F_i)$, summed over the components F_i of F. If M^3 is a compact orientable 3-manifold, define for $\mu \in H^1(M^3)$

$$\|\mu\| = \min\{\chi_-(F) \mid F \subset M^3 \text{ is an embedded surface dual to } \mu\} \, .$$

Thurston shows that this is the restriction to $H^1(M^3)$ of a continuous piecewise linear pseudonorm $\|-\| : H^1(M^3; \mathbf{R}) \to \mathbf{R}_+$.

If $L(\underline{m})$ is a multilink on $L = (\Sigma, \mathcal{K})$, we call $\|\underline{m}\|$ *the norm of the multilink* $L(\underline{m})$. If $F \subset \Sigma_0$ is a surface dual to $\underline{m}$, the procedure above to change F into a Seifert surface cannot increase $\chi_-(F)$. Hence $\|\underline{m}\| = \min\{\chi_-(F) | F$ is a Seifert surface for $L(\underline{m})\}$. Thus norm plays the role for multilinks which the usual genus plays for knots. A Seifert surface F for $L(\underline{m})$ which minimizes $\chi_-(F)$, and which has no closed components is called a *minimal Seifert surface.*

THEOREM 3.3. *Suppose* $L(\underline{m}) = L'(\underline{m}') \underset{S' \quad S''}{\overline{\hspace{4em}}} L''(\underline{m}'')$ *and neither* S' *nor* S'' *is an unknot summand of the respective link* L' *or* L''. *Then a minimal Seifert surface for* $L(\underline{m})$ *can be obtained by pasting minimal Seifert surfaces for* $L'(\underline{m}')$ *and* $L''(\underline{m}'')$, *so* $\|\underline{m}\| = \|\underline{m}'\| + \|\underline{m}''\|$.

REMARK. If one of S' and S'' is an unknot summand, then $L(\underline{m})$ is the disjoint sum of the multilinks obtained by deleting S' and S'' from $L'(\underline{m}')$ and $L''(\underline{m}'')$. It is easily verified that norm is also additive for disjoint sum of multilinks (in fact for connected sum of any 3-manifolds).

Proof of 3.3. Let Σ_0, Σ'_0, Σ''_0 be the respective link exteriors, so $\Sigma_0 = \Sigma'_0 \cup \Sigma''_0$, pasted along a torus T. Our knottedness assumption on S' and S'' is, by Dehn's lemma [Pa], precisely the condition that T be incompressible in Σ_0.

If $F' \subset \Sigma_0'$ and $F'' \subset \Sigma_0''$ are minimal Seifert surfaces for $L'(\underline{m}')$ and $L''(\underline{m}'')$, then their intersections with T are both dual to the restriction of $\underline{m}$ to T, so after an isotopy we may assume $F' \cap T = F'' \cap T$. Thus $F = F' \cup F''$ is a Seifert surface for $L(\underline{m})$. Now $\chi_-(F) = \chi_-(F') + \chi_-(F'')$, since no component of F' or F'' which meets T is a disk, by incompressibility of T. Hence $\|\underline{m}\| \le \chi_-(F) = \chi_-(F') + \chi_-(F'') = \|\underline{m}'\| + \|\underline{m}''\|$. It remains to show $\|\underline{m}\| \ge \|\underline{m}'\| + \|\underline{m}''\|$.

Let F be a minimal Seifert surface for $L(\underline{m})$. We may assume, by isotoping F if necessary, that F is transverse to T. Let $F_0' = F \cap \Sigma_0'$, $F_0'' = F \cap \Sigma_0''$, and let F' and F'' be Seifert surfaces for $L'(\underline{m}')$ and $L''(\underline{m}'')$ constructed from F_0' and F_0'' as in the proof of Lemma 3.1. Now $\chi_-(F_0') + \chi_-(F_0'')$ exceeds $\chi_-(F)$ by the number of new discs created by cutting F into the pieces F_0' and F_0''. Any such disk meets T in an inessential circle, by incompressibility of T, and so disks are spanned back into the corresponding boundary components of F_0' and F_0'' in constructing F' and F''. This cancels the excess of $\chi_-(F_0') + \chi_-(F_0'')$ over $\chi_-(F)$ again, so $\chi_-(F') + \chi_-(F'') \le \chi_-(F) = \|\underline{m}\|$. Thus $\|\underline{m}'\| + \|\underline{m}''\| \le \|\underline{m}\|$, completing the proof.

COROLLARY 3.4. *The norm of an irreducible multilink* $L(\underline{m})$ *is the sum of the norms of its splice components.*

REMARK. Our proof above of additivity of norm applies with no change to cutting any compact 3-manifold along incompressible tori, and a completely parallel proof applies to cutting along incompressible annuli. This shows:

PROPOSITION 3.5. *If* M *is a compact oriented 3-manifold and* $\mu \in H^1(M)$, *then the Thurston norm* $\|\mu\|$ *of* μ *is the sum of the norms of the restrictions of* μ *to the Jaco-Shalen-Johannson components of the connected sum components of* M.

4. *Fibration and monodromy*

The multilink $L(\underline{m}) = (\Sigma, m_1 S_1, \cup \cdots \cup m_n S_n)$ is said to be a fibered multilink if there exists a fibration $\Lambda : \Sigma_0 \to S^1$ of the link exterior Σ_0, all of whose fibers are Seifert surfaces for $L(\underline{m})$. Equivalently, the requirement is that if we identify $H^1(\Sigma_0) = [\Sigma_0, S^1]$, the group of homotopy classes of maps of Σ_0 to S^1, then $\underline{m} \in [\Sigma_0, S^1]$ contains a fibration Λ.

We shall discuss fiberability of $L(\underline{m})$ in terms of the splice decomposition and describe the monodromy of the fibration in the fibered case. Again the results generalize to arbitrary 3-manifolds.

We remark that the fibration for $L(\underline{m})$, if it exists, is unique up to isotopy. This is a general fact for compact 3-manifolds and is a special case of a much stronger result of Blank and Laudenbach on uniqueness up to isotopy of nonzero closed 1-forms representing given cohomology classes. We need for this a relative version of their result [B-L], which, Blank informs us, follows by the same methods. One can also prove the uniqueness of fibration by standard 3-manifold techniques. We sketch this. Let $\Lambda, \Lambda' : M^3 \to S^1$ be homotopic fibrations. Then their fibers F and F' are homeomorphic, since the corresponding infinite cyclic cover $\overline{M}^3$ of M^3 is homeomorphic to both $F \times \mathbf{R}$ and $F' \times \mathbf{R}$. Also the monodromy maps $F \to F$ and $F' \to F'$ are given by a covering transformation of $\overline{M}^3$ and are hence homotopic after identifying F with F' by this homeomorphism. By surface theory they are then even isotopic (see for instance [Z-V-C]). Thus the corresponding mapping tori are homeomorphic, that is, there exists a homeomorphism $M \to M$ taking Λ to Λ'. By construction this homeomorphism is homotopic to the identity, hence isotopic to the identity by Waldhausen [Wa].

PROPOSITION 4.1. *If* $L(\underline{m})$ *is a fibered multilink, then any fiber is a minimal Seifert surface and conversely, any minimal Seifert surface is isotopic to a fiber.*

Again, this proposition generalizes to any 3-manifold fibered over S^1, as will be clear from the proof.

Proof. Let Σ_0 be the link exterior, and $\overline{\Sigma}_0$ the infinite cyclic cover of Σ_0 classified by $\underline{m}$. The fibration $\Lambda : \Sigma_0 \to S^1$ leads to a product structure $\overline{\Sigma}_0 \cong F \times \mathbf{R}$, where F is a fiber of Λ. Let $F' \subset \overline{\Sigma}_0$ be any minimal Seifert surface lifted from Σ_0 to $\overline{\Sigma}_0$. Projecting $\overline{\Sigma}_0 = F \times \mathbf{R}$ to F gives a mapping $j : F' \to F$ which is a homeomorphism near the boundary. By minimality of F' and a theorem of Kneser [Kn] (see also [Ed]), f is homotopic, relative boundary, to a homeomorphism. By Stallings [St] then F' is a fiber of some fibration $\Lambda' : \Sigma_0 \to S^1$. Since Λ' also represents $\underline{m}$, this completes the proof.

THEOREM 4.2. *The multilink* $L(\underline{m})$ *is fibered if and only if it is irreducible and each of its multilink splice components is fibered.*

More generally a map Λ *of a compact connected oriented 3-manifold* M^3 *to* S^1 *is homotopic to a fibration if and only if either* M^3 *is the total space of an* S^2 *or* T^2 *bundle over* S^1 *and* Λ *is essential, or* M *is irreducible and the restriction of* Λ *to each Jaco-Shalen-Johannson component of* M^3 *is homotopic to a fibration.*

Proof. If $\Lambda : M \to S^1$ is a fibration with fiber $F \neq S^2$, then $\widetilde{M} = \widetilde{F} \times \mathbf{R}$ has no essential embedded 2-spheres, so M is irreducible. Moreover, the boundary components of M are all tori, so in the Jaco-Shalen-Johannson splitting M is cut only along incompressible tori. We must show that, after an isotopy, these tori can be assumed to be transverse to all fibers of Λ, so Λ restricts to a fibration on each piece of the splitting. This is true so long as $F \neq T^2$ by the theorem of Thurston [Th 3] and Roussarie [Ro], which says that an incompressible torus in a foliated 3-manifold with no genus 1 leaves can be isotoped to be transverse to the foliation. It is not hard also to give a group theoretic argument, using Stalling's fibration theorem [St].

Conversely, if $\Lambda : M^3 \to S^1$ is a map which can be homotoped to a fibration on each piece of some splitting of M along tori, then, since homotopic fibrations $T^2 \to S^1$ are isotopic, the fibrations of the pieces of M can be isotoped to match up along the tori to produce a fibration $\Lambda': M \to S^1$. We claim this can be done so that Λ' is homotopic to Λ. In fact, this holds automatically if every torus of the decomposition separates M (true for a link exterior), since then $H^1(M) \to H^1(M')$ is injective, where M' is M split along the tori. In general, if Λ' is not homotopic to Λ, one can correct this by doing suitable "twists" at the separating tori before pasting; we leave the details to the reader.

We leave to the reader also the verification that any essential map of the total space of an S^2 or T^2 bundle over S^1 to S^1 is homotopic to a fibration, which completes the proof of the theorem.

There is an alternate approach to the above theorem which gives slightly more information. If $\Lambda: M^3 \to S^1$ is a fibration with fiber F then the monodromy $h: F \to F$ of this fibration is a well-defined diffeomorphism up to isotopy. By Thurston's theory of surface diffeomorphisms [F-L-P], we can, after an isotopy of h, find a closed 1-manifold $S \subset F$ with $h(S) = S$, such that F cut open along S consists of pieces F_i with $h|F_i$ either of finite order or pseudo-Anosov, up to isotopy, for each i, and such that no proper submanifold of S has this property. This splitting is unique up to isotopy. The mapping torus M of h splits along tori into the mapping tori M_i of $h|F_i$, which will be Seifert or simple according as $h|F_i$ is isomorphic to a finite order or pseudo-Anasov map. This splitting of M is minimal, since if Seifert structures matched up across some torus of this splitting, the corresponding part of $S \subset F$ would have been redundant. This shows:

ADDENDUM TO 4.2. *If* $\Lambda: M^3 \to S^1$ *is a fibration with monodromy* $h: F \to F$, *then the Jaco-Shalen-Johannson splitting of* M *into Seifert and simple pieces is induced by Thurston's splitting of* (F, h) *into periodic and pseudo-Anosov pieces.*

Let $L(\underline{m})$ be a fibered multilink with monodromy $h: F \to F$. Denote by $\Delta_i \in \mathbf{Z}[t]$ the characteristic polynomial of the algebraic monodromy

$H_i(h): H_i(F) \to H_i(F)$. Thus $\Delta_0 = t^d - 1$, where d is the number of components of F. This d is simply the greatest integer divisor of $\underline{m}$, as is clear for instance from the exact homotopy sequence

$$\cdots \to \pi_1(\Sigma_0) \xrightarrow{(\underline{m})_\#} \pi_1(S^1) \to \pi_0(F) \to 0 \text{ for the fibration } \Lambda : \Sigma_0 \to S^1 .$$

Denote

$$\Delta_* = \Delta_1/\Delta_0 \in \mathbf{Q}(t)$$

THEOREM 4.3. *Δ_* for $\mathbf{L}(\underline{m})$ is the product of the Δ_* for the splice components of $\mathbf{L}(\underline{m})$.*

Proof. More generally if $h : F \to F$ is any homeomorphism of a "nice" space (e.g. finite complex) we can define

$$\Delta_*(h) = \prod_i \Delta_i(h)^{((-1)^{i-1})} = \Delta_0^{-1}\Delta_1\Delta_2^{-1}\Delta_3 \cdots$$

where $\Delta_i(h) \in \mathbf{Z}[t]$ is the characteristic polynomial of $H_i(h)$. If $F = F_1 \cup F_2$ and $h(F_1) \subset F_1$ and $h(F_2) \subset F_2$, the Mayer Vietoris sequence implies

$$\Delta_*(h) = \Delta_*(h|F_1)\Delta_*(h|F_2)/\Delta_*(h|F_1 \cap F_2) .$$

If (F, h) is the monodromy of a fibered multilink obtained by splicing two multilinks with fibers F_1 and F_2, then $F_1 \cap F_2$ is a collection of circles permuted by h, so $\Delta_*(h|F_1 \cap F_2) = 1$. Thus the theorem follows.

A different proof and generalization of this theorem are in the next section.

5. *Alexander polynomial*

We recall the definition of the *Alexander polynomial* $\Delta^{\mathbf{L}}(t_1, \cdots, t_n)$ of a link $\mathbf{L} = (\Sigma, S_1 \cup \cdots \cup S_n)$. Let $\pi : \tilde{\Sigma}_0 \to \Sigma_0$ be the universal abelian cover of the link exterior and let $\tilde{p} = \pi^{-1}(p)$ be a typical fiber. The group of covering transformations is $H_1(\Sigma_0) = \mathbf{Z}^n$. Writing t_i for the covering transformation associated to an oriented meridian of S_i, the group $H_1(\tilde{\Sigma}_0, \tilde{p})$ becomes a module over $\mathbf{Z}[t_1, t_1^{-1}, \cdots, t_n, t_n^{-1}]$. The i-th fitting ideal $F_i = F_i(H_1(\tilde{\Sigma}_0, \tilde{p}))$ is the ideal in $\mathbf{Z}[t_1, t_1^{-1}, \cdots, t_n, t_n^{-1}]$ generated by the $(g-i) \times (g-i)$ minors of a presentation matrix for $H_1(\tilde{\Sigma}_0, \tilde{p})$, where g is the number of generators. $\Delta^{\mathbf{L}}(t_1, \cdots, t_n)$ is the greatest common divisor of F_1. We shall show later that $F_1 = \Delta_*^{\mathbf{L}} \cdot I$ for some $\Delta_*^{\mathbf{L}} \in \mathbf{Z}[t_1, t_1^{-1}, \cdots, t_n, t_n^{-1}]$, where $I = (t_1 - 1, \cdots, t_n - 1)$ is the augmentation ideal. Thus $\Delta^{\mathbf{L}} = \Delta_*^{\mathbf{L}}$ if $n \geq 2$ and $\Delta^{\mathbf{L}} = (t_1 - 1)\Delta_*^{\mathbf{L}}$ if $n = 1$.

Note that $\Delta^{\mathbf{L}}$ is only well defined up to multiplication by units, that is elements $\pm t_1^{a_1} \cdots t_n^{a_n} \in \mathbf{Z}[t_1, t_1^{-1}, \cdots, t_n, t_n^{-1}]$.

If $\mathbf{L}(\underline{m}) = (\Sigma, m_1 S_1 \cup \cdots \cup m_n S_n)$ is a multilink on $\mathbf{L}$, then the *Alexander polynomial* $\Delta^{\mathbf{L}(\underline{m})} \in \mathbf{Z}[t, t^{-1}]$ of $\mathbf{L}(\underline{m})$ is defined just like the several variable Alexander polynomial, but using the infinite cyclic cover $\tilde{\Sigma}_0(\underline{m}) \to \Sigma_0$ determined by $\underline{m}$ instead of the universal abelian cover. It is well defined up to multiplication by units $\pm t^i$. Define $\Delta_*^{\mathbf{L}(\underline{m})} = (t^d - 1)^{-1} \Delta^{\mathbf{L}(\underline{m})}$, where $d = \gcd(m_1, \cdots, m_n)$. The following result shows that $\Delta_*^{\mathbf{L}(\underline{m})}$ is still a polynomial if $n \geq 2$.

This proposition is a mild generalization of a lemma of Milnor [Mi 2, Lemma 10.1].

PROPOSITION 5.1.

i) $\Delta_*^{\mathbf{L}(\underline{m})} = \Delta_*^{\mathbf{L}}(t^{m_1}, \cdots, t^{m_n})$. That is

$$\Delta^{\mathbf{L}(\underline{m})} = \begin{cases} (t^d - 1)\Delta^{\mathbf{L}}(t^{m_1}, \cdots, t^{m_n}) & \text{if } n \geq 2 . \\ \Delta^{\mathbf{L}}(t^{m_1}) & \text{if } n = 1 . \end{cases}$$

ii) *The order ideal* $F_0 H_1(\tilde{\Sigma}_0(\underline{m}))$ *of the* $\mathbf{Z}[t, t^{-1}]$*-module* $H_1(\tilde{\Sigma}_0(\underline{m}))$ *is the principal ideal generated by* $\Delta^{\mathbf{L}(\underline{m})}$.

iii) *If* $L(\underline{m})$ *is fibered with monodromy* $h : F \to F$, *then* $\Delta^{L(\underline{m})} = \Delta_1$, *the characteristic polynomial of the algebraic monodromy* $H_1(h) : H_1(F) \to H_1(F)$, *up to a factor* $\pm t^i$.

Proof. The proofs are as in Milnor (loc. cit.) once we know that $F_1H_1(\tilde{\Sigma}_0, \tilde{p}) = \Delta_* \cdot I$ with $\Delta_* \in Z[t_1, t_1^{-1}, \cdots, t_n, t_n^{-1}]$ and $I = (t_1-1, \cdots, t_n-1)$. Namely, denote $\tilde{\Sigma}_0(\underline{m})$ by $\bar{\Sigma}_0$ for short and let $\bar{p}$ be a fiber of $\bar{\Sigma}_0 \to \Sigma_0$. Then $H_1(\bar{\Sigma}_0, \bar{p}) = H_1(\tilde{\Sigma}_0, \tilde{p}) \otimes_R Z[t, t^{-1}]$, where $R = Z[t_1, t_1^{-1}, \cdots, t_n, t_n^{-1}]$ and $Z[t, t^{-1}]$ has the R-module structure given by $t_i \to t^{m_i}$. Thus

$$F_1H_1(\bar{\Sigma}_0, \bar{p}) = \Delta_*(t^{m_1}, \cdots, t^{m_n})(t^{m_1}-1, \cdots, t^{m_n}-1) .$$

But a simple induction shows

$$(t^{m_1}-1, \cdots, t^{m_n}-1) = (t^d - 1) ,$$

so i) follows, and moreover $F_1H_1(\bar{\Sigma}_0, \bar{p})$ is the principal ideal $(\Delta^{L(\underline{m})})$. Now the exact sequence

$$0 \to H_1(\bar{\Sigma}_0) \to H_1(\bar{\Sigma}_0, \bar{p}) \to H_0(\bar{p}) \to H_0(\bar{\Sigma}_0) \to 0$$

implies that

$$H_1(\bar{\Sigma}_0, \bar{p}) \cong H_1(\bar{\Sigma}_0) \oplus Z[t, t^{-1}] \text{ as a } Z[t, t^{-1}]\text{-module} .$$

Hence $F_1H_1(\bar{\Sigma}_0, \bar{p}) \cong F_0H_1(\bar{\Sigma}_0)$, proving ii). Finally, in the fibered case $H_1(\bar{\Sigma}_0) = H_1(F)$, so iii) follows.

The fact that $F_1H_1(\tilde{\Sigma}_0, \tilde{p})$ does indeed have the form $\Delta_* \cdot I$ is usually proven, for links in S^3, from a Wirthinger presentation of the group of the link, a method not applicable here. It is in fact true in much greater generality than we need here, and is of independent interest. We therefore postpone the general statement and proof to the next section.

Part ii) of the above proposition, which is often used as a definition of the single variable Alexander polynomial, makes easy the computation of $\Delta^{L(\underline{m})}$ in terms of the splice decomposition.

THEOREM 5.2. *Suppose* $L(\underline{m}) = (\Sigma, m_1 S_1 \cup \cdots \cup m_n S_n)$ *is the result of splicing*

$$L'(\underline{m}') = (\Sigma', m'_0 S'_0 \cup m_1 S_1 \cup \cdots \cup m_k S_k)$$

and

$$L''(\underline{m}'') = (\Sigma'', m''_0 S''_0 \cup m_{k+1} S_{k+1} \cup \cdots \cup m_n S_n)$$

along S'_0, S''_0, *with* $0 \le k < n$. *Then*

$$\Delta_*^{L(\underline{m})} = \Delta_*^{L'(\underline{m}')} \Delta_*^{L''(\underline{m}'')}$$

unless $m'_0 = m''_0 = 0$ *and* $k = 0$, *in which case*

$$\Delta_*^{L(\underline{m})} = \Delta_*^{L''_0(\underline{m})}, \text{ where } L''_0(\underline{m}) = (\Sigma'', m_1 S_1 \cup \cdots \cup m_n S_n).$$

Proof. Let Σ_0, Σ'_0, Σ''_0 be the relevant link exteriors, so $\Sigma_0 = \Sigma'_0 \cup_T \Sigma''_0$ pasted along a torus T. Let $\overline{\Sigma}_0 = \overline{\Sigma}'_0 \cup_{\overline{T}} \overline{\Sigma}''_0$ be the infinite cyclic cover classified by $\underline{m}$.

We first note that if $H_2(\overline{\Sigma}_0) \neq 0$ then $\Delta^{L(\underline{m})} = 0$. Indeed, Σ_0, as a connected 3-manifold with boundary, has the homotopy type of a 2-complex so $H_2(\overline{\Sigma}_0)$ is a free $\mathbf{Z}[t,t^{-1}]$ module. Now if $H_2(\overline{\Sigma}_0) \neq 0$, let $X \subset \overline{\Sigma}_0$ be an embedded surface representing a nontrivial homology class. X cannot separate $\overline{\Sigma}_0$, since then it would be homologous to $t^N X$ for some large N. Thus we can find a 1-cycle S having nontrivial intersection number with X. This S must generate a free $\mathbf{Z}[t,t^{-1}]$ submodule of $H_1(\overline{\Sigma}_0)$, so $\Delta^{L(\underline{m})} = 0$.

We first prove the theorem assuming $m'_0 \neq 0$ or $m''_0 \neq 0$. Let $m = \gcd(m'_0, m''_0)$. Then $\overline{T}$ consists of m components, each homeomorphic to $S^1 \times \mathbf{R}$. Consider the Mayer-Vietoris sequence

$$\cdots \to H_2(\bar{\Sigma}'_0) \oplus H_2(\bar{\Sigma}''_0) \to H_2(\bar{\Sigma}_0) \xrightarrow{\partial} H_1(\bar{T}) \to H_1(\bar{\Sigma}'_0) \oplus H_1(\bar{\Sigma}''_0) \to \cdots .$$

Now if $H_2(\bar{\Sigma}_0)$ is non-trivial then it is a free $\mathbf{Z}[t, t^{-1}]$-module, so $\ker(\partial)$ is non-trivial, so at least one of $H_2(\bar{\Sigma}'_0)$ or $H_2(\bar{\Sigma}''_0)$ is non-trivial. In this case both sides of the equation to be proved are zero and we are done. Thus assume $H_2(\bar{\Sigma}_0) = 0$. Then the Mayer Vietoris sequence becomes:

$$0 \to H_1(\bar{T}) \to H_1(\bar{\Sigma}'_0) \oplus H_1(\bar{\Sigma}''_0) \to H_1(\bar{\Sigma}_0) \to H_0(\bar{T}) \to$$

$$H_0(\bar{\Sigma}'_0) \oplus H_0(\bar{\Sigma}''_0) \to H_0(\bar{\Sigma}_0) \to 0 .$$

Equating the products of the orders over $\mathbf{Z}[t, t^{-1}]$ of alternate terms of this sequence gives

$$[t^m - 1][\Delta^{L(\underline{m})}][(t^{d'} - 1)(t^{d''} - 1)]$$

$$= [\Delta^{L'(\underline{m}')} . \Delta^{L''(\underline{m}'')}][t^m - 1][t^d - 1] ,$$

which simplifies to the equation to be proved.

It remains to discuss the case $m'_0 = m''_0 = 0$. We first note that $\Delta^{L'(\underline{m}')} = 0$. Indeed $\bar{T} \subset \partial \bar{\Sigma}'_0$ and $\bar{T}$ consists of infinitely many tori. Now by Poincare duality, any two elements of $\mathrm{Ker}\,(H_1(\bar{T}) \to H_1(\bar{\Sigma}'_0))$ have zero intersection number (this is also clear geometrically). It follows that $\mathrm{Im}\,(H_1(\bar{T}) \to H_1(\bar{\Sigma}'_0))$ is still infinitely generated over $\mathbf{Z}$, so $\Delta^{L'(\underline{m}')} = 0$. The same of course holds for $\Delta^{L''(\underline{m}'')}$. Now if $k > 0$ then $\bar{T}$ represents non-trivial homology in $H_2(\bar{\Sigma}_0)$, so by an earlier comment, $\Delta^{L(\underline{m})} = 0$ and we are done. If $k = 0$, then the link exteriors for

$$L(\underline{m}) = (\Sigma, m_1 S_1 \cup \cdots \cup m_n S_n)$$

and

$$L''_0(\underline{m}) = (\Sigma'', m_1 S_1 \cup \cdots \cup m_n S_n)$$

differ just in the replacement of the homology solid torus $\Sigma'_0 \subset \Sigma_0$ by a genuine solid torus. Moreover this homology solid torus and genuine solid torus are trivially covered in the relevant infinite cyclic covers of the link exteriors, so these infinite cyclic covers have identical homology, implying the final case of the theorem.

As a corollary of the above results, we can also describe the behavior of the several variable Alexander polynomial under splicing. If $L = (\Sigma, S_1 \cup \cdots \cup S_n)$, recall our notation

$$\Delta_*^L = \begin{cases} \Delta^L(t_1, \cdots, t_n) & \text{if } n > 1 \\ (t_1 - 1)^{-1} \Delta^L(t_1) & \text{if } n = 1 \end{cases}$$

THEOREM 5.3. *Let* $L = (\Sigma, S_1 \cup \cdots \cup S_n)$ *be the result of splicing* $L' = (\Sigma', S_0' \cup S_1 \cup \cdots \cup S_k)$ *and* $L'' = (\Sigma'', S_0'' \cup S_{k+1} \cup \cdots \cup S_n)$ *along* S_0', S_0'', *with* $0 \le k < n$. *Let* $b_i = \ell(S_0', S_i)$ *for* $i = 1, \cdots, k$ *and* $a_j = \ell(S_0'', S_j)$ *for* $j = k+1, \cdots, n$. *Then, unless* $k = a_1 = \cdots = a_n = 0$,

$$\Delta_*^L = \Delta_*^{L'}(T_0', t_1, \cdots, t_k) \cdot \Delta_*^{L''}(T_0'', t_{k+1}, \cdots, t_n), \text{ with}$$

$$T_0' = t_{k+1}^{a_{k+1}} \cdots t_n^{a_n} \text{ and } T_0'' = t_1^{b_1} \cdots t_k^{b_k}.$$

If $k = a_1 = \cdots = a_n = 0$ *then* $\Delta_*^k = \Delta_*^{L_0''}$, *where* L_0'' *is* L'' *with* S_0'' *deleted.*

Proof. By 5.1 i) and 5.2 the claimed equation is valid after any specialization $(t_1, \cdots, t_n) = (t^{m_1}, \cdots, t^{m_n})$. This implies its validity as it stands.

6. *Fitting ideals associated to Abelian covers*

Recall that a group or module has *defect* ≥ 1 if it admits a presentation with strictly more generators then relations. All homology groups in the rest of this section will have $\mathbf{Z}$-coefficients.

THEOREM 6.1. *Let* X *be a path-connected space and let* $\widetilde{X}$ *be a connected covering with finitely generated abelian covering group* A *of torsion free rank* ≥ 2. *If* $\widetilde{p}$ *is a fiber of the covering map* $\widetilde{X} \to X$, *then* $H_1(\widetilde{X}, \widetilde{p})$, *regarded as a* $\mathbf{Z}[A]$ *module, has first fitting ideal of the form*

$$F_1(H_1(\widetilde{X}, \widetilde{p})) = IJ ,$$

where I *is the augmentation ideal of* $Z[A]$ *and* $J \subset Z[A]$ *is another ideal. If* $\pi_1(X)$ *has defect* ≥ 1, *then so does* $H_1(\widetilde{X}, \widetilde{p})$ *and* J *may be taken to be principal.*

REMARK. Given the existence of an ideal J satisfying the theorem, the result will in general be valid for many different ideals J, the largest and most banal of which will be the "ideal quotient" $(I : F_1(H_1(\widetilde{X}, \widetilde{p})))$. A choice depending more subtly on the covering is described in Theorem 6.3, and is the (unique) choice which is principal in the defect ≥ 1 case.

Theorem 6.1 may be applied to link exteriors, to give the result needed in the previous section.

LEMMA 6.2. *Let* M *be a compact 3-manifold with* $\partial M \neq \emptyset$.

a) *If* $\chi(\partial M) = 0$, *then* $\chi(M) = 0$.

b) *If* $\chi(M) = 0$ *and* M *is connected, then* $\Pi_1(M)$ *has defect* 1.

Proof of Lemma 6.2.

a) The closed 3-manifold $M \underset{\partial M}{\cup} (-M)$ has Euler characteristic $2\chi(M) - \chi(\partial M) = 2\chi(M)$ by additivity; but its Euler characteristic is also 0 by Poincaré duality.

b) M is homotopic to a wedge of, say, n circles to which, say, m 2-cells have been attached. Since $0 = \chi(M) = 1 - n + m$, we get $m = n - 1$. But $\pi_1(M)$ has a presentation with generators corresponding to the circles and relations corresponding to the 2-cells.

To prove Theorem 6.1 we use a general algebraic result. Recall that an ideal I in a commutative ring has *depth* ≥ 2 if there exist $x_1, x_2 \in I$

with x_1 a non-zerodivisor on R and x_2 a non-zerodivisor on R/x_1. For example, if A is a finitely generated abelian group, and $g_1, g_2 \in A$ generate a rank 2 free abelian subgroup, then $x_1 = 1 - g_1$, $x_2 = 1 - g_2$ satisfy the above property in the group ring $Z[A]$, so the augmentation ideal of A has depth ≥ 2.

THEOREM 6.3. *Let* M *be a module over a commutative noetherian ring* R. *If there exists an epimorphism* $M \to I$ *from* M *to an ideal of depth* ≥ 2, *then* $F_1(M) = IJ$ *for some ideal* J, *which may be taken to be principal if* M *has defect* ≥ 1. *The ideal* J *may be determined uniquely by the following requirement:*

Suppose that M *is generated by elements* $m_1, \cdots, m_n$ *whose images in* $I \subseteq R$ *are* $a_1, \cdots, a_n$. *If* $R^m \overset{\phi}{\to} R^n \to M$ *is a presentation of* M *in terms of these generators, then* J *admits a set of generators* $\mathfrak{F}_K$, *indexed by the* (n-1)*-element subsets* K *of a basis for* R^m, *such that* $a_i \mathfrak{F}_k$ *is the minor of* ϕ *corresponding to the basis elements* K *of* R^m *and all but the* i^{th} *basis element of* R^n.

Theorem 6.1 was proved, for abelian coverings of classical link exteriors, by Torres [To] using the properties of the Wirthinger presentation of π_1. Special cases (at least) of Theorems 6.1 and 6.3 may be found implicitly in work of Hillman [H, Theorem IV]. Theorem 6.3 may be regarded as a result on the Steinitz-Fox-Smythe "row and column" invariants of a module. A general theory of such invariants, in a modern setting, may be found in [B-E], and the proof we give below in an easy application of that theory. For an elementary exposition, see [No].

We are grateful to J. Levine for pointing out to us the relevance of "defect ≥ 1" in Lemma 6.2 and in the proof of Theorem 6.1.

Proof of Theorem 6.1. The long exact sequence in homology (with Z-coefficients) of the pair $(\widetilde{X}, \widetilde{p})$ yields

$$0 \to H_1(\widetilde{X}) \to H_1(\widetilde{X}, \widetilde{p}) \to H_0(\widetilde{p}) \to H_0(\widetilde{X}) \to 0 \, ,$$
$$\begin{array}{ccc} \| & & \| \\ Z[A] & \to & Z \end{array}$$

and thus a map of $H_1(\widetilde{X}, \widetilde{p})$ onto the augmentation ideal I of $Z[A]$.

Suppose that $\pi_1(X)$ has defect ≥ 1. The second part of 6.1 will follow if we show that $H_1(\widetilde{X}, \widetilde{p})$ has defect ≥ 1 as a $Z[A]$-module.

Since $H_1(\widetilde{X}, \widetilde{p})$ (and even the class of the extension of $Z[A]$-modules $0 \to H_1(\widetilde{X}) \to H_1(\widetilde{X}, \widetilde{p}) \to I \to 0$) depends only on the map $\pi_1(X) \to A$ corresponding to the covering $\widetilde{X}$, we may suppose that, for some n, X is obtained by attaching $< n$ 2-cells to a wedge X_1 of n circles, with base point p, say, corresponding to the relations and generators in a given defect ≥ 1 presentation of $\pi_1(X)$. If we write $\widetilde{p}$ and $\widetilde{X}_1$ for the preimages of p and X_1 in $\widetilde{X}$, the exact homology sequence of the triple $(\widetilde{X}, \widetilde{X}_1, \widetilde{p})$ yields

$$H_2(\widetilde{X}, \widetilde{X}_1) \to H_1(\widetilde{X}_1, \widetilde{p}) \to H_1(\widetilde{X}, \widetilde{p}) \to H_1(\widetilde{X}, \widetilde{X}_1) = 0 \, .$$

Now $H_1(\widetilde{X}_1, \widetilde{p})$ is a free $Z[A]$-module on lifts of the n circles of X_1, while $H_2(\widetilde{X}, \widetilde{X}_1)$ is generated as an A-module by lifts of the $< n$ 2-cells of X, so the above sequence is a defect ≥ 1 presentation of $H_1(\widetilde{X}, \widetilde{p})$.

Proof of Theorem 6.3. Let

$$R^m \xrightarrow{\phi_2} R^n \xrightarrow{\phi_1} R$$

be a complex of free R-modules such that

$$\text{Coker } \phi_2 = M \, ,$$

$$\text{Im } \phi_1 = I \, ,$$

and ϕ_1 induces the given epimorphism $M \to I$.

Let $\phi_2' : R^{m'} \to R^n$ be a map from a free R-module onto $\ker \phi_1 \subseteq R^n$. Since $\text{Im } \phi_2 \subset \ker \phi_1$, there exists a map ψ making the diagram

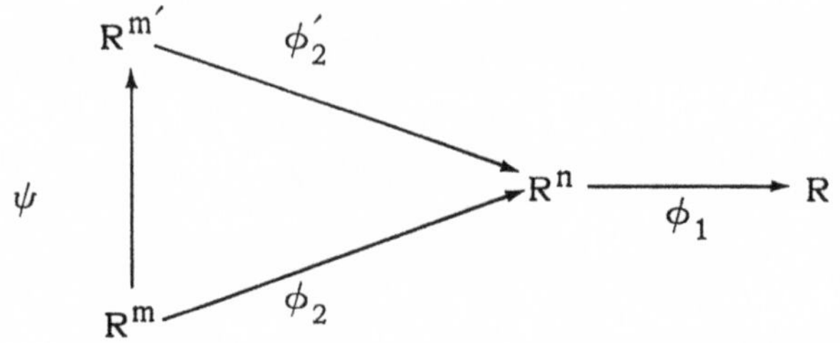

commute.

For any map $F \xrightarrow{\phi} G$ we define $\operatorname{rk}\phi$ to be the largest k with $\overset{k}{\Lambda}\phi = 0$, and we write $I(\phi)$ for the ideal of minors of ϕ of order $= \operatorname{rk}\phi$ (see [B-E]). If we invert a non-zerodivisor in I, then $\ker\phi_1$ becomes free of rank $n-1$ and we deduce that $\operatorname{rk}\phi_2' = n-1$, and $I(\phi_2')$ contains a non-zerodivisor (in fact depth $I(\phi_2') \geq 2$). On the other hand, $I(\phi_1) = I$ has depth ≥ 2 by the hypothesis so the sequence

$$0 \longrightarrow R \xrightarrow[\phi_1^*]{} R^{n^*} \xrightarrow[\phi_2'^*]{} R^{m'^*}$$

obtained by dualizing is exact [B-E]. The "First Structure Theorem" of [B-E] now yields the existence and uniqueness of a map α making the diagram:

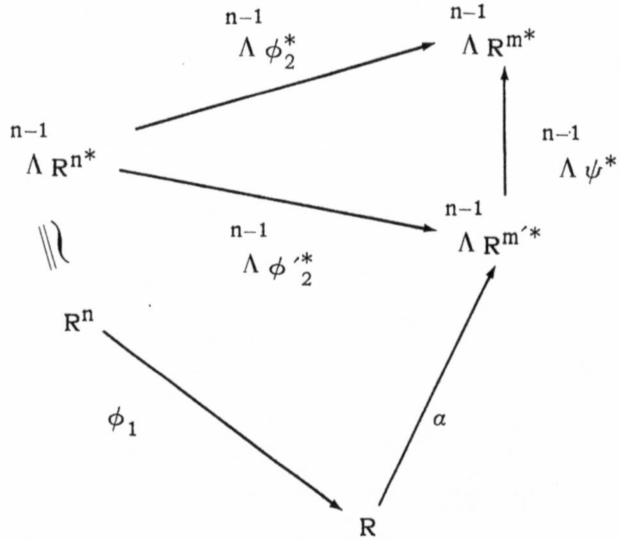

commute. The ideal

$$J = I(\overset{n-1}{\Lambda}\ \psi^* \cdot \alpha)\ ,$$

which may also be described as $\mathrm{Im}(\alpha^* \cdot \overset{n-1}{\Lambda}\ \psi)$, satisfies the condition of the theorem. If M has defect ≥ 1, and we choose the presentation $R^m \underset{\phi_2}{\longrightarrow} R^n$ of M with $m < n$, then $\overset{n-1}{\Lambda}\ R^{m^*} \cong R$ or 0, so J is principal.

Appendix to Chapter I
ALGEBRAIC LINKS

In this appendix we describe how a splice diagram for an algebraic link may be derived from Puiseux expansions. We also describe a simple method for deriving the power-series equation satisfied by a Puiseux expansion.

Algebraic links

Let $f(x,y)$ be a complex polynomial vanishing at 0, and let

$$C = \{(x,y) \in \mathbb{C}^2 | f(x,y) = 0\}$$

be the corresponding algebraic plane curve. For all sufficiently small $\varepsilon > 0$, the 3-sphere

$$S^2_\varepsilon = \{(x,y) \in \mathbb{C}^2 | \ |(x,y)| = \varepsilon\}$$

meets C transversely in a link, which has a natural orientation coming from that of C. An oriented link (S^3,K) obtained in this way is said to be an *algebraic link.*

To describe (S^3,K) more fully we may (changing variables if necessary) solve $f(x,y) = 0$ for y in terms of x, obtaining a set of solutions which are fractional power series, called *Puiseux series*, in x. Each fractional power series solution gives rise to a *branch* of the curve, and thus to one component of the link (two solutions which differ by a change of variable of the form $x \mapsto \zeta x$, with ζ a root of unity, may describe the same branch). It is not hard to show that all but finitely many terms of the power series can be removed without changing the topology of the link.

The resulting minimal Puiseux series are usually written in the form

$$y = c_1 x^{m_1/n_1} + c_2 x^{m_2/n_1 n_2} + \cdots$$

with $0 \neq c_i \in \mathbf{C}$, $m_1/n_1 < m_2/n_1 n_2 < \cdots$,

and each pair (m_i,n_i) relatively prime, and then (m_i,n_i) are called the *Puiseux Pairs* for the corresponding branch. For us, however, it will be more convenient to write the solutions in the multiplicative form:

$$*) \qquad y = x^{q_1/p_1}(a_1 + x^{q_2/p_1 p_2}(a_2 + \cdots (a_{s-1} + x^{q_s/p_1 \cdots p_s}(a_s + \cdots) \cdots)))$$

with $p_i, q_i > 0$ and (p_i,q_i) relatively prime for all i. Expanding the product as a power series, one sees at once that the (p_i,q_i) are determined from the (m_i,n_i) by the formulas $p_i = n_i$, $q_1 = m_1$, $q_i = m_i - m_{i-1} n_i$ for $i > 1$. The pairs (p_i,q_i) might well be called the "Newton pairs" of the expansion, since they are the numbers which are produced by the direct application of Newtons method for computing the expansion ([Wa]).

Making a suitable change of variables we may assume that, in each of the expansions corresponding to C, we have $|y| << |x|$ for $|x|$ small. We may then replace the intersection of C with S^3_ε by the intersection of C with the solid torus

$$R = \{(x,y) \in \mathbf{C}^2 \mid |x| = \varepsilon, |y| \leq \varepsilon\},$$

which is naturally embedded as an unknotted solid torus in the (topological) sphere

$$R \cup \{(x,y) \in \mathbf{C}^2 \mid |x| \leq \varepsilon, |y| = \varepsilon\}.$$

Isotopy arguments [K-N] show that the link $C \cap R$, embedded in S^3 as above, is the same as the link $C \cap S^3_\varepsilon$ in S^3_ε (the change to this point of view seems to be due to Kähler).

To understand this link, consider first the case of a knot, given, say, by the expansion *), in R. A first approximation to it is the p_1, q_1 torus knot, (S^3, K_1), given by

$$y = a_1 x^{q_1/p_1}$$

in $R \subset S^3$.

To see that this really is the (p_1,q_1) torus knot, set $x = \varepsilon t^{\phi}$, where t runs once around the complex unit circle $S^1 \subset C$. Then y is a constant times t^{q_1} so (x,y) runs p_1 times around in the longitudinal direction in R (the x-axis) while running q_1 times around the meridianal direction of R (the y-axis). We will represent (S^3, K_1) by the graph

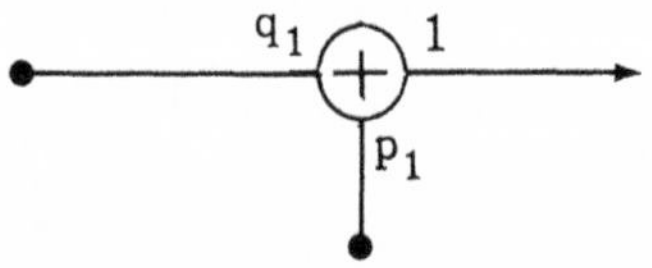

(such graphs are introduced systematically in section 8).

As a second approximation to the knot K, consider the knot K_2 given by

$$y = x^{q_1/p_1}(a_1 + a_2 x^{q_2/p_1 p_2}).$$

Changing our parametrization to $x = \varepsilon t^{p_1 p_2}$, we see that since ε is very small K_2 will lie in a small tubular neighborhood of K_1, and will in fact be a cable on K_1. Clearly, it will follow K_1 around p_2 times in a longitudinal direction, so for some integer α_2 it will be a (p_2, α_2) cable on K_1. If we let $L_2 = (S^3, K_0 \cup K_0(p_2, \alpha_2))$ be the link consisting of an unknotted circle K_0 and then (p_2, α_2)-cable on it, then by Proposition 1.1 we see that

$$(S^3, K_2) = (S^3, K_1) \underset{K_1 \quad K_0}{\overline{\qquad\qquad}} L_2 .$$

We will represent K_2 by the diagram

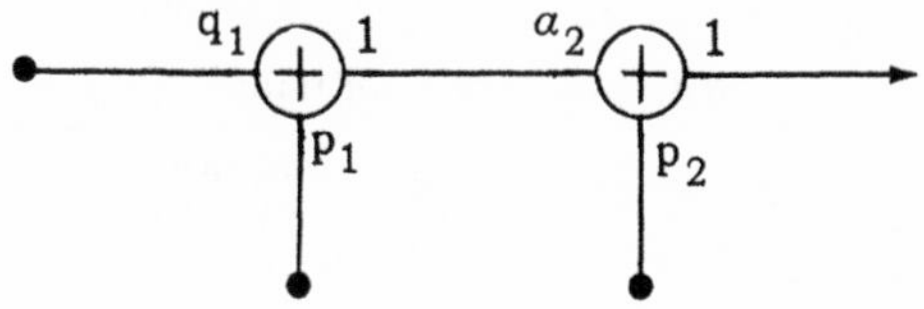

which should be regarded as a "composite" of the diagrams

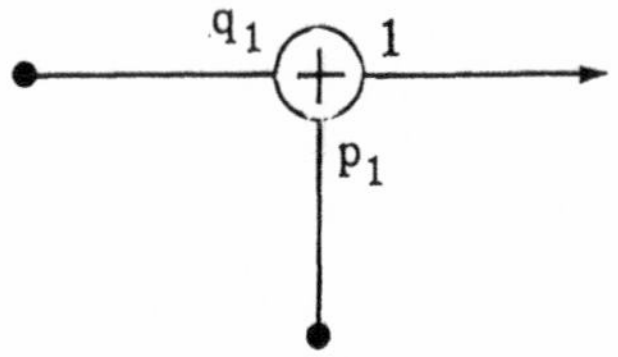

and

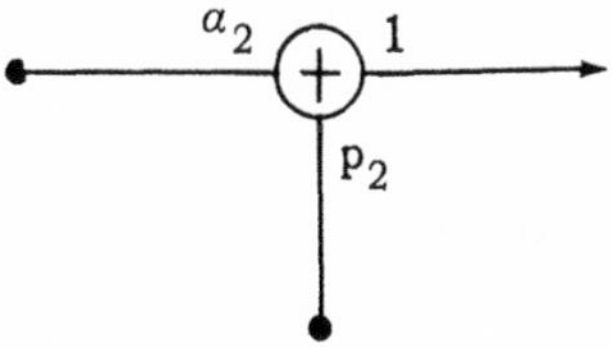

for (S^3, K_1) and L_2, respectively.

Repeating these constructions inductively, we see that the knot K represented by *) is, for suitable integers $\alpha_1 = q_1$, $\alpha_2, \cdots, \alpha_s$, represented as the (p_s, α_s) cable on the (p_{s-1}, α_{s-1}) cable on the ... (p_1, α_1)-cable on the unknot, and we represent it by the diagram

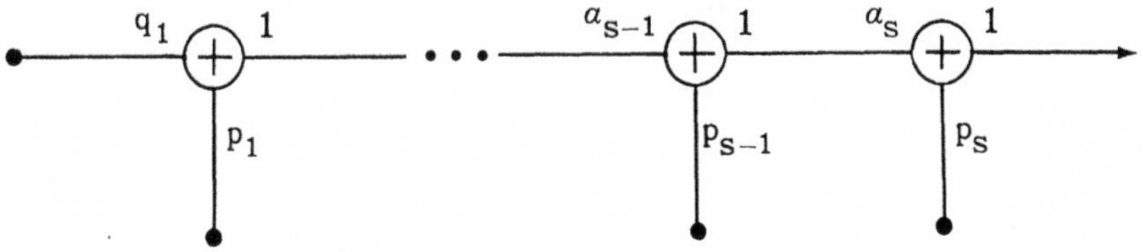

It remains to determine the α_i.

PROPOSITION 1A.1. *The* α_i *above are given from the Newton pairs* (p_i, q_i) *by the formulas*

$$\alpha_1 = q_1$$

and, for $i \geq 1$,

$$\alpha_{i+1} = q_{i+1} + p_i p_{i+1} \alpha_i .$$

Note that the *topological pair* (p_i, α_i) is relatively prime, since (p_i, q_i) is, and that the "algebraicity condition" $p_i, q_i > 0$ becomes $p_i > 0$, $\alpha_{i+1} > p_i p_{i+1} \alpha_i$ in terms of the topological pairs. (See section 9A for an extension of this.)

Proof. The fact that $\alpha_1 = q$ has already been noted. To prove the second formula, consider the knots K_{i-1}, K_i and K_{i+1}, where K_j is the knot in $R \subset S^3$ corresponding to the parametrization

$$y = x^{q_1/p_1}(\cdots(a_{j-1} + a_j x^{q_j/p_1 \cdots p_j})\cdots)$$

in R. If $\varepsilon = |x|$ is sufficiently small, then we may choose tubular neighborhoods $N(K_{i-1})$ and $N(K_i)$ such that $N(K_i)$ is contained in the interior of $N(K_{i-1}) - K_{i-1}$ and K_{i+1} is contained in the interior of $N(K_i)$. We will write M_j, L_j for the topologically standard meridian and longitude of K_j in $\partial N(K_j)$ for $j = i-1, i$.

It suffices to show that, in the homology of $N(K_i) - K_i$, we have $K_{i+1} \sim p_{i+1} L_i + (q_{i+1} + p_i p_{i+1} \alpha_i) M_i$. Let $L \subset N(K_{i-1}) - K_i$ be the knot obtained from K_i by moving each point of K_i a certain small distance in the direction "directly away" from K_{i-1}. We could describe L analytically by a parametrization of the form

$$y = x^{q_1/p_1}(\cdots(a_{i-1} + x^{q_i/p_1 \cdots p_i}(a_i + \delta))\cdots),$$

for suitably small real δ, and it follows from this description that, in the homology of $N(K_i) - K_i$,

$$K_{i+1} \sim p_{i+1} L + q_{i+1} M_i,$$

so it suffices to show that $L \sim L_i + p_i \alpha_i M_i$.

Since L is obviously homologous to K_i in $N(K_i)$, it suffices, by the definition of L_i, to show that $L - p_i \alpha_i M_i$ is null-homologous in $S^3 - K_i$. We will do this by showing that, in the homology of $N(K_{i-1}) - (K_{i-1} \cup K_i)$,

$$L - p_i \alpha_i M_i \sim p_i L_{i-1} \,,$$

or, equivalently, $L \sim p_i L_{i-1} + p_i \alpha_i M_i$.

Now in the homology of $N(K_{i-1}) - (K_{i-1} \cup K_i)$ we clearly have $p_i M_i \sim M_{i-1}$, so the desired formula becomes

$$L \sim p_i L_{i-1} + \alpha_i M_{i-1} \,.$$

On the other hand, from the construction of L, we see that we could continue to "push" away from K_{i-1} until reaching $\partial N(K_{i-1})$ without encountering K_i, so L is at least homologous to some linear combination of L_{i-1} and M_{i-1}. Since, in the homology of $N(K_{i-1}) - K_{i-1}$ we have $L \sim K_i \sim p_i L_{i-1} + \alpha_i M_{i-1}$, this concludes the proof.

In describing algebraic links with several components we may restrict ourselves to the case of 2 components, since the general case offers no new difficulties.

Suppose, then, that $(S^3, K \cup K')$ is an algebraic link with two components, corresponding to the two distinct developments

$$K : \; y = x^{q_1/p_1}(a_1 + \cdots (a_{s-1} + a_s x^{q_s/p_1 \cdots p_s}) \cdots)$$

$$K' : \; y = x^{q_1'/p_1'}(a_1' + \cdots (a_{r-1}' + a_r' x^{q_r'/p_1' \cdots p_r'}) \cdots) \,.$$

Suppose, moreover, that n is the number of common terms; that is, $(p_1, q_1) = (p_1', q_1'), \cdots, (p_n, q_n) = (p_n', q_n')$ and $a_1 = a_1', \cdots, a_n = a_n'$ but $r = n$ or $s = n$ or $(p_{n+1}, q_{n+1}) \neq (p_{n+1}', q_{n+1}')$ or $a_{n+1} \neq a_{n+1}'$. There are several cases to consider:

First, suppose $r = n$ (the case $s = n$ is analogous). Adapting the notation for the successive approximations to K used above (so that $K = K_s$) we have $K' = K_n$. We represent this link by the diagram:

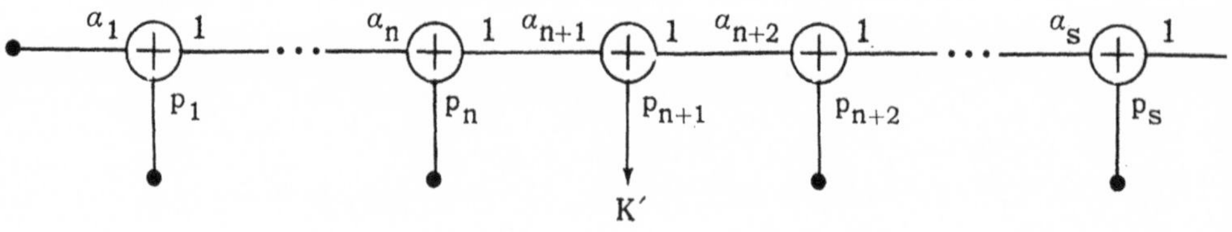

We may now suppose that r and s are both $> n$. Let K_n, K_{n+1} be the nth and $(n+1)^{st}$ approximants to K as above, and let K'_{n+1} be the $(n+1)^{st}$ approximant to K′, in the analogous way. Both K_{n+1} and K'_{n+1} are cables on K_n.

Suppose that $q_{n+1}/p_{n+1} < q'_{n+1}/p'_{n+1}$. Then, since $|x|$ is very small, we will have,

$$|a_{n+1}x^{q_{n+1}/p_1 \cdots p_{n+1}}| > |a'_{n+1}x^{q'_{n+1}/p'_1 \cdots p'_{n+1}}| ,$$

so that K'_{n+1} is a cable on the boundary of a *smaller* tubular neighborhood of K_n then is K_{n+1}.

Since the next cabling operations in the construction of K and K′ take place in small tubular neighborhoods of K_{n+1} and K'_{n+1}, they do not interfere, so that with (p_i, α_i) derived from (p_i, q_i) and (p'_i, α'_i) derived from (p'_i, q'_i) as above, we may represent the link $(S^3, K \cup K')$ by the graph

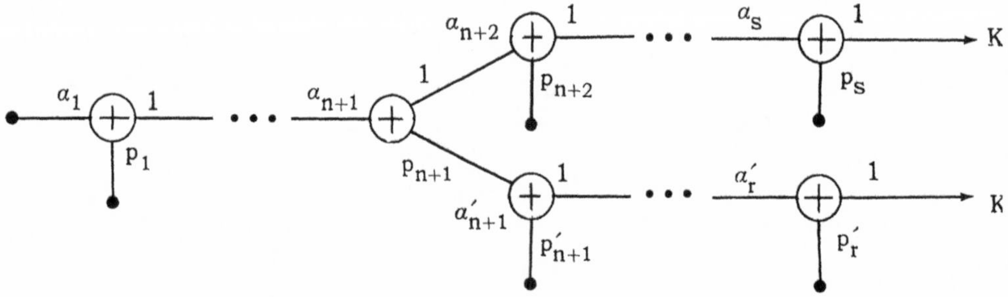

Finally, suppose $q_{n+1}/p_{n+1} = q'_{n+1}/p'_{n+1}$, so that $(p_{n+1}, q_{n+1}) = (p'_{n+1}, q'_{n+1})$ but $a_{n+1} \neq a'_{n+1}$. In this case, K_{n+1} and K'_{n+1} are both

(p_{n+1}, α_{n+1})-cables on K_n, and after a topologically harmless change of a_{n+1} (say) we could assume that K'_{n+1} is formed on a smaller torus than K_n, and proceed as above. However, our notation allows a more symmetrical graph, as well. We may represent $(S^3, K \cup K')$ in this case by the graph:

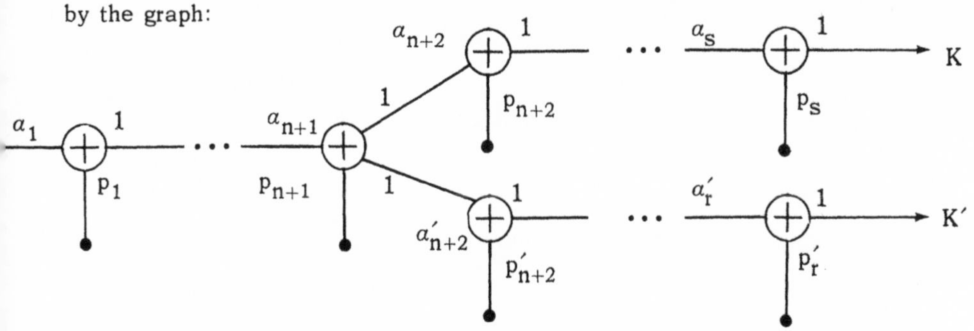

This concludes our description of algebraic links via graphs.

A final remark on the "non-reduced" case is in order. The branches of C also correspond to distinct irreducible factors of f as a (convergent or formal) power series. If a factor of f occurs with multiplicity >1, so that f is not "reduced", then it is natural to consider the corresponding branch of C to have multiplicity >1 (this will be its natural scheme structure) and to consider the corresponding component of the algebraic link $C \subset R \subset S^3$ to have multiplicity >1 (again, this will happen naturally if we treat $C \subset R$ as a real-analytic space instead of a set). The notion of a "multiple component of a link" was given corresponding topological significance above in section 3. We introduced there the terminology "multilink" for a link with an integer multiplicity assigned to each component. We also described how, by taking linking numbers, such an assignment of multiplicities can also be interpreted as a linear functional on the first homology of the link complement Σ_0, hence as a cohomology class $\underline{m} \in H^1(\Sigma_0; Z)$.

The equation of a plane branch

We present a simple method for computing an irreducible power series equation $f(x,y) = 0$ satisfied by a given Puiseux expansion

$$*)\quad \begin{cases} x = t^n \\ y = a_1 t^{m_1} + a_2 t^{m_2} + \cdots \\ a_i \neq 0,\ m_1 \geq n,\ \mathrm{GCD}(n, m_1, m_2, \cdots) = 1 . \end{cases}$$

Of course the equation for a link with several branches is obtained by multiplying the equations of the individual branches.

The method is standard in the theory of integral extensions, but seems not widely noticed by those dealing with singularities. Everything here can be adapted to ground fields of arbitrary characteristic, but we leave this to the reader.

We write Id_n for an $n \times n$ identity matrix.

PROPOSITION. *Let* $x, y \in \mathbb{C}[[t]]$ *be such that* $\mathbb{C}[[t]]$ *is the integral closure of* $\mathbb{C}[[x,y]]$ *(or equivalently,* t *is a rational function of* x *and* y *). If* $u_0, \cdots, u_{n-1} \in \mathbb{C}[[t]]$ *reduces to a basis of* $\mathbb{C}[[t]]/(x)$ *, then there exist unique elements* $v_{ij} \in \mathbb{C}[[x]]$ *such that*

$$yu_i = \sum_{j=0}^{n-1} v_{ij} u_j \qquad i = 0, \cdots, n-1 .$$

If we set

$$f(x,u) = \det(u \cdot \mathrm{Id}_n - (v_{ij})) ,$$

then $f(x,y) \equiv 0$ *and* $f(x,u)$ *is irreducible over the field of Laurent series* $\mathbb{C}((x))$*. In particular,* $f(x,u) = 0$ *defines the same branch of a plane curve as* $x(t), y(t)$*.*

REMARK. If x and y are convergent power series and the u_i are chosen similarly (they can always be taken to be $1, t, \cdots, t^{n-1}$ if one wishes), then the v_{ij} and f will be convergent.

In the situation of (*), and choosing the u_i as $1, t, \cdots, t^{n-1}$, if y is a polynomial then the v_{ij} and f will be polynomials.

Proof. By Nakayama's lemma, $\mathbf{C}[\![t]\!]/(x^p)$ is a free $\mathbf{C}[\![x]\!]/(x^p)$-module, with basis the classes of the u_i, for every p. Since $\mathbf{C}[\![x]\!]$ is complete it follows that $\mathbf{C}[\![t]\!]$ is itself a free $\mathbf{C}[\![x]\!]$ module with basis the u_i. (In the convergent case the corresponding statement holds because $\mathbf{C}\{x\}$ is Henselian.) This implies the first statement of the proposition.

The element $y \in \mathbf{C}[\![t]\!]$ acts $\mathbf{C}[\![x]\!]$-linearly, by multiplication, on $\mathbf{C}[\![t]\!]$ and $f(x,u)$ is its characteristic polynomial, so $f(x,y) \equiv 0$. To prove irreducibility, it suffices to note that the quotient field of $\mathbf{C}[\![x]\!][y]$ is $\mathbf{C}((t))$, and $[\mathbf{C}((t)) : \mathbf{C}((x))] = n$, the degree of f in y.

If x and y have the form given by (*), we can be much more explicit. Namely if we choose $u_j = t^i$ $(j = 0, \cdots, n-1)$ we obtain:

COROLLARY. *Suppose* x *and* y *are as in* (*), *and rewrite* y *as*

$$y(t) = \sum_{\ell \geq 0} b_\ell t^\ell .$$

For $s = 0, \cdots, n-1$ *set*

$$v_s = - \sum_{\ell \equiv s(n)} b_\ell x^{(\ell - s)/n} ,$$

$$f(x,u) = \det \begin{pmatrix} u+v_0 & v_1 & v_2 & \cdots & v_{n-2} & v_{n-1} \\ xv_{n-1} & u+v_0 & v_1 & v_2 & \cdots & v_{n-2} \\ xv_{n-2} & xv_{n-1} & \ddots & \ddots & \ddots & \vdots \\ \vdots & xv_{n-2} & \ddots & \ddots & \ddots & v_2 \\ xv_2 & \vdots & \ddots & \ddots & \ddots & v_1 \\ xv_1 & xv_2 & \cdots & xv_{n-2} & xv_{n-1} & u+v_0 \end{pmatrix}$$

then f *is irreducible over* $\mathbb{C}((x))$ *and* $f(x,y) \equiv 0$.

EXAMPLE. Let

$$x = t^4$$

$$y = t^6 + at^7 .$$

This is the simplest Puiseux series with more than one characteristic pair. We get

$$f(x,u) = \det \begin{pmatrix} u & 0 & -x & -ax \\ -ax^2 & u & 0 & x \\ -x^2 & -ax^2 & u & 0 \\ 0 & -x^2 & -ax^2 & u \end{pmatrix}$$

$$= u^4 - 2x^3u^2 - 4a^2x^5u + x^6 - a^4x^7 .$$

This is the example (with $a=1$) given in part 1 of the introduction.

Chapter II
CLASSIFICATION

7. *Classification of Seifert links*

Recall that a Seifert link is a link $L = (\Sigma, K) = (\Sigma', S_1 \cup \cdots \cup S_n)$ whose exterior $\Sigma_0 = \Sigma' - \text{int } N(K)$ admits a Seifert fibration.

LEMMA 7.1. *An irreducible link* $L = (\Sigma, K)$ *is a Seifert link if and only if* K *is an invariant set for some effective* S^1*-action on* Σ. *Moreover, this action on* Σ *can be chosen fixed point free unless* L *is the following link in* S^3 :

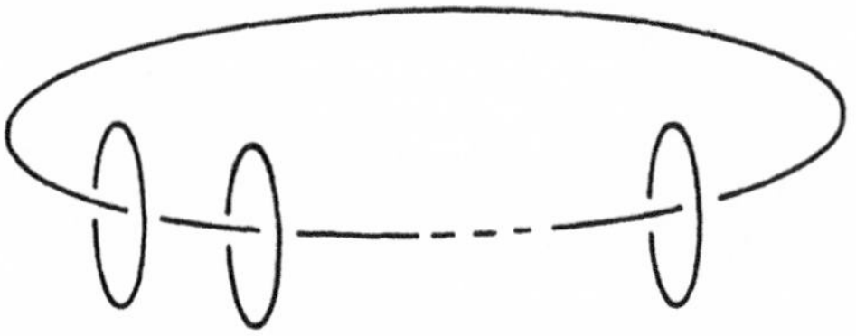

Proof. Suppose L is a Seifert link and let $\pi: \Sigma_0 \to F$ be the Seifert fibration. If F were non-orientable, its orientation cover would induce a nontrivial double covering of Σ_0, trivial along $\partial \Sigma_0$, and hence also a nontrivial double covering of Σ. This is impossible, since Σ is a homology sphere. Thus, F is orientable, so the fibers of π can be consistently oriented, so π is the orbit map $\Sigma_0 \to \Sigma_0/S^1 = F$ of an effective fixed point free S^1-action on Σ_0. This action can be extended over Σ so that each S_i is either an orbit or a component of the fixed point set.

Conversely, suppose K is invariant under an effective S^1-action on Σ. This action induces a Seifert fibration of the link exterior unless it has fixed points there. By Orlik and Raymond [O-R] the only effective S^1-action with fixed points on a homology sphere is the standard S^1-action on S^3 with fixed point set $S^1 \subset S^3$. A union of $n > 1$ orbits of this action is the disjoint sum of m unknotted circles in S^3, hence not an irreducible link. Thus in this case K must include the fixed point set and thus be as pictured in the lemma.

The above lemma says that, except in the case described, Σ itself is Seifert fibered and K is a collection of fibers. We must thus first recall Seifert's classification [Se] of Seifert fibered homology spheres (see also [N-R] and [Ne 2]). Let $\alpha_1, \cdots, \alpha_n$ be pairwise coprime integers with $n \geq 3$ and each $\alpha_i \geq 2$. Then there exists a unique Seifert fibered 3-manifold whose (unnormalized, as in [Ne 1] and [N-R]) Seifert invariant has the form

$$(0:(\alpha_1, \beta_1), \cdots, (\alpha_n, \beta_n))$$

with

$$\sum_{i=1}^{n} \beta_i \alpha_1 \cdots \hat{\alpha}_i \cdots \alpha_n = 1 . \tag{7.2}$$

This manifold is denoted $\Sigma(\alpha_1, \cdots, \alpha_n)$; it is an homology sphere. To any oriented Seifert fibered homology sphere $\Sigma \not\cong S^3$, there is a unique unordered n-tuple $(\alpha_1, \cdots, \alpha_n)$ as above and a unique orientation sign $\pm$, such that $\Sigma \cong \pm \Sigma(\alpha_1, \cdots, \alpha_n)$ (orientation preserving homeomorphism). This homeomorphism may be chosen to preserve Seifert fibered structure.

We shall make use of two explicit descriptions of $\Sigma(\alpha_1, \cdots, \alpha_n)$. The first is the standard topological construction of Seifert manifolds:

TOPOLOGICAL DESCRIPTION. Let $F_0 = S^2 - \text{int}(D_1^2 \cup \cdots \cup D_n^2)$ be the n-fold punctured 2-sphere. $\Sigma = \Sigma(\alpha_1, \cdots, \alpha_n)$ can be characterized as a closed oriented 3-manifold in which we can embed disjoint solid tori $(D^2 \times S^1)_1, \cdots, (D^2 \times S^1)_n$ such that:

a). *If* $\Sigma_0 = \Sigma - \text{int}((D^2 \times S^1)_1 \cup \cdots \cup (D^2 \times S^1)_n)$ *then there exists a fibration* $\pi: \Sigma_0 \to F_0$ *with fiber* S^1, *so* $\Sigma_0 \cong F_0 \times S^1$.

b). *If* $R \subset \Sigma_0$ *is a section to* π *and for* $i = 1, \cdots, n$

$$\begin{aligned} Q_i &= -\partial R \cap (D^2 \times S^1)_i \\ H_i &= \textit{typical fiber of } \pi \textit{ in } \partial (D^2 \times S^1)_i \end{aligned}$$

then $\alpha_i Q_i + \beta_i H_i$ *is nullhomologous in* $(D^2 \times S^1)_i$, *where* $\beta_1, \cdots, \beta_n$ *satisfy equation* (7.2).

Note that (7.2) determines $\beta_i \pmod{\alpha_i}$ and forces $\gcd(\alpha_i, \beta_i) = 1$, but does not uniquely determine the β_i. This fact corresponds, in our geometric description, to the fact that the Q_i, and hence the β_i, depend on the choice of the section $R \subset \Sigma_0$. The description can be reversed to give an explicit construction of Σ: paste solid tori $(D^2 \times S^1)_i$ into $F_0 \times S^1$ in such a way that the homology class

$$\alpha_i (S^1 \times \{1\}) + \beta_i (\{1\} \times S^1)$$

in the i-th boundary component of $F_0 \times S^1$ is nullhomologous in $(D^2 \times S^1)_i$ after pasting.

The S^1-bundle structure on Σ_0 extends to a Seifert fibered structure on Σ with the core circle S_i of $(D^2 \times S^1)_i$ as a singular fiber of degree α_i. Thus $(\Sigma(\alpha_1, \cdots, \alpha_n), S_1 \cup \cdots \cup S_n)$ is a special case of a Seifert link. But of course a general Seifert link may not include all the singular fibers as link components and may include some nonsingular fibers. To obtain a uniform notation for arbitrary Seifert links it is useful to generalize this discussion a bit.

We required $n \geq 3$ and $\alpha_i \geq 2$ in the definition of $\Sigma(\alpha_1, \cdots, \alpha_n)$ only to ensure a one to one classification of Seifert fiberable homology spheres. The above description of $\Sigma(\alpha_1, \cdots, \alpha_n)$ applies as written to *any* n-tuple $(\alpha_1, \cdots, \alpha_n)$ of pairwise coprime integers and yields a homology sphere. Define S_i to be the core circle of $(D^2 \times S^1)_i$, oriented so that $\alpha_i Q_i + \beta_i H_i$ represents an oriented meridian, that is

$$\ell(S_i, \alpha_i Q_i + \beta_i H_i) = 1 .$$

Then $(\Sigma(\alpha_1, \cdots, \alpha_n), S_1 \cup \cdots \cup S_n)$ is a well-defined Seifert link. We show later (remark after Lemma 7.5) that this orientation of S_i is the same as its orientation as a fiber of the Seifert structure on Σ if α_i is positive.

Negative α_i's lead only to changes in orientation and α_i's equal to 1 correspond to non-singular fibers. This is described in detail in Proposition 7.3 below, but first we give the second description of these links.

ANALYTIC DESCRIPTION. Let $A = (a_{ij})$ be any $(n-2) \times n$ matrix, all of whose maximal minors are nonzero. Assume $n \geq 2$ and $\alpha_i \geq 1$ for $i = 1, \cdots, n$. Define

$$V_A(\alpha_1, \cdots, \alpha_n) = \{Z \in \mathbf{C}^n | a_{i1} Z_1^{\alpha_1} + \cdots + a_{in} Z_n^{\alpha_n} = 0,\ i = 1, \cdots, n-2\}.$$

Then in [Ne 2] (see also [N-R]) it is shown that

$$\Sigma(\alpha_1, \cdots, \alpha_n) \cong V_A(\alpha_1, \cdots, \alpha_n) \cap S^{2n-1},$$

with Seifert fibration induced by the S^1-action

$$t(Z_1, \cdots, Z_n) = (t^{q_1} Z_1, \cdots, t^{q_n} Z_n)$$

$$t \in S^1 = \{t \in \mathbf{C} | \ \|t\| = 1\},\quad q_i = \alpha_1 \cdots \alpha_n / \alpha_i .$$

Moreover, the link component S_i is the intersection of $\Sigma = \Sigma(\alpha_1, \cdots, \alpha_n)$ with the hyperplane $Z_i = 0$.

This analytic description applies also to $(\alpha_1, \cdots, \alpha_n) = (0, 1, \cdots, 1)$ by using

$$V_A(0, 1, \cdots, 1) = \{Z \in \mathbf{C}^n | a_{i2} Z_2 + \cdots + a_{in} Z_n = 0,\ i = 1, \cdots, n-2\}$$

and letting S^1 act by

$$t(Z_1, \cdots, Z_n) = (tZ_1, Z_2, \cdots, Z_n).$$

PROPOSITION 7.3.

1) *Every Seifert link is isomorphic to one of the form*

$$(\pm \Sigma(\alpha_1, \cdots, \alpha_n), \pm S_1 \cup \cdots \cup \pm S_m)$$

with $\alpha_i \geq 0$ *and* $m \leq n$.

2) *Seifert links are invertible, that is*

$$(\Sigma(\alpha_1,\cdots,\alpha_n), S_1 \cup \cdots \cup S_n) \cong (\Sigma(\alpha_1,\cdots,\alpha_n), -S_1 \cup \cdots \cup -S_n).$$

3) $(\Sigma(-\alpha_1,\alpha_2,\cdots,\alpha_n), S_1 \cup \cdots \cup S_n) \cong (-\Sigma(\alpha_1,\cdots,\alpha_n), -S_1 \cup S_2 \cup \cdots \cup S_n)$.

4) $(\Sigma(\alpha_1,\cdots,\alpha_k,1,\cdots,1), S_1 \cup \cdots \cup S_n)$ *is obtained from* $(\Sigma(\alpha_1,\cdots,\alpha_k), S_1 \cup \cdots \cup S_k)$ *by adding* $n-k$ *nonsingular fibers* $S_{k+1},\cdots,S_n$.

5) $(\Sigma(\alpha_1,\alpha_2,1,\cdots,1), S_1 \cup S_2 \cup \cdots \cup S_n) \cong (S^3, S_1 \cup S_2 \cup (n-2)\, S_1(\alpha_1,\alpha_2))$ *as pictured below*

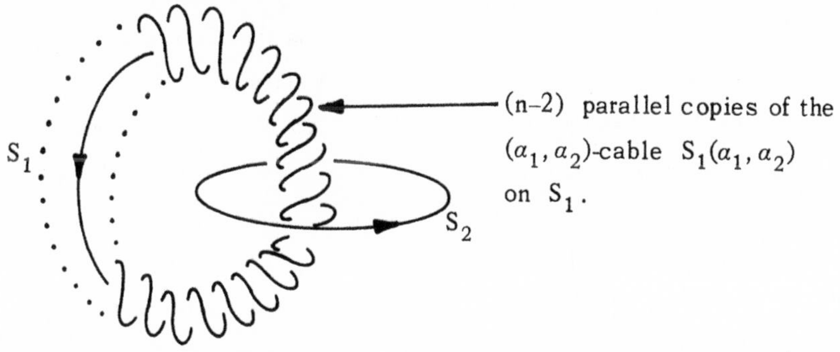

6) *In particular* $(\Sigma(0,1,\cdots,1), S_1 \cup \cdots \cup S_n)$ *is the link in* S^3:

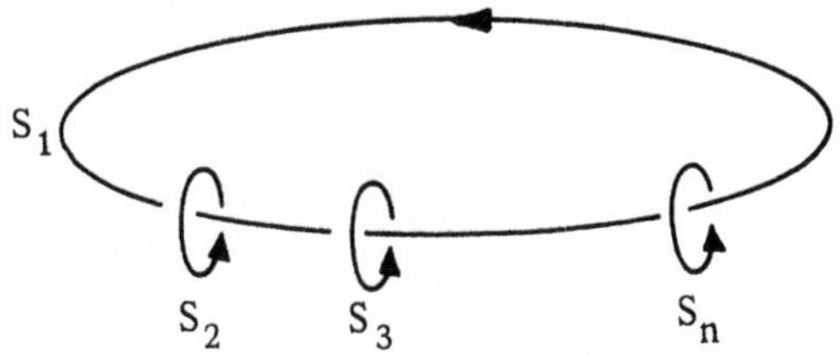

Proofs.

1) Properly we should prove this after parts 4) to 5). Assuming them, the fact that we have all Seifert links (Σ, K) if $\Sigma \neq S^3$ is clear from our

earlier discussion plus part 4) of this theorem. For $\Sigma = S^3$ it follows from the fact that every circle action on S^3 is equivalent to one of the ones described in our proof of parts 5) and 6) below, see for instance [O-R], or [Se].

2) Any Seifert manifold admits an orientation preserving self-diffeomorphism mapping any desired finite collection of fibers to themselves with reversed orientation. In fact, this diffeomorphism can always be chosen as a fiber preserving involution, see Montesinos; [Mn] or Bonahon and Siebenmann [B-S], where these involutions are discussed in detail. That is, Seifert links are even "strongly invertible."

3) If $((\alpha_1,\beta_1),\cdots,(\alpha_n,\beta_n))$ satisfies equation (7.2) for $(\alpha_1,\cdots,\alpha_n)$, then $((-\alpha_1,\beta_1),(\alpha_2,-\beta_2),\cdots,(\alpha_n,-\beta_n))$ satisfies it for $(-\alpha_1,\alpha_2,\cdots,\alpha_n)$. Thus, from the construction of $(\Sigma(-\alpha_1,\alpha_2,\cdots,\alpha_n),S_1\cup\cdots\cup S_n)$ we see it can be obtained from $(\Sigma(\alpha_1,\cdots,\alpha_n),S_1\cup\cdots\cup S_n)$ by reversing the fiber orientation (which reverses the orientation of $\Sigma(\alpha_1,\cdots,\alpha_n)$) and then reversing the orientation of S_1.

4) If $((\alpha_1,\beta_1),\cdots,(\alpha_k,\beta_k))$ is a Seifert invariant for $\Sigma(\alpha_1,\cdots,\alpha_k)$ then $((\alpha_1,\beta_1),\cdots,(\alpha_k,\beta_k),(1,0),\cdots,(1,0))$ satisfies equation (7.2) for $(\alpha_1,\cdots,\alpha_k,1,\cdots,1)$. But $(1,0)$ Seifert pairs correspond to tubular neighborhoods of non-singular fibers, so 4) follows.

5) By directly checking Seifert invariants, or by using the analytic description of $\Sigma(\alpha_1,\cdots,\alpha_n)$ plus part 2) of this theorem, one verifies that $\Sigma(\alpha_1,\alpha_2)$, for $\alpha_1 \neq 0$ and $\alpha_2 \neq 0$, is the Seifert fibering of

$$S^3 = \{(Z_1,Z_2) \in C^2 \mid |Z_1|^2 + |Z_2|^2 = 1\}$$

by orbits of the S^1-action

$$t(Z_1,Z_2) = (t^{\alpha_2}Z_1, t^{\alpha_1}Z_2), \quad t \in S^1 = \{t \in C \mid |t| = 1\}.$$

The two orbits $S_1 = \{Z_1 = 0\}$ and $S_2 = \{Z_2 = 0\}$ form the standard link

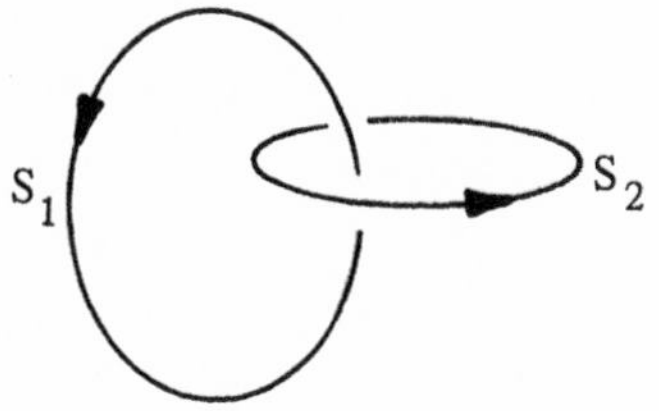

in S^3 and all other orbits are (a_1, a_2)-cables on S_1. The case that one of a_1 and a_2 is zero, say $a_1 = 0$, follows the same way except that S_1 is the fixed point set of the S^1 action.

PROPOSITION 7.4. *In the link* $(\Sigma(a_1, \cdots, a_n), S_1 \cup \cdots \cup S_n)$ *the linking number* $\ell(S_i, S_j)$ *is* $a_1 \cdots \hat{a}_i \cdots \hat{a}_j \cdots a_n$.

Proof. We shall give two proofs. The first uses the topological description of $\Sigma = \Sigma(a_1, \cdots, a_n)$. In the notation of that construction,

$$\Sigma = \Sigma_0 \cup (D^2 \times S^1)_1 \cup \cdots \cup (D^2 \times S^1)_n$$

and Σ_0 is a genuine S^1-bundle. Moreover, if $R \subset \Sigma_0$ is a section and $Q_j = -\partial R \cap (D^2 \times S^1)_j$, we have $a_j Q_j + \beta_j H = 0$ in $H_1((D^2 \times S^1)_j; \mathbf{Z})$. Here H denotes the homology class of a nonsingular fiber, wherever it occurs.

Let M_j and L_j be a standard meridian and longitude for the component S_j, considered as simple curves in $\partial(D^2 \times S^1)_j$.

LEMMA 7.5. *The bases* $\{M_j, L_j\}$ *and* $\{Q_j, H\}$ *for* $H_1(\partial(D^2 \times S^1)_j)$ *are related by*

$$\begin{pmatrix} M_j \\ L_j \end{pmatrix} = \begin{pmatrix} a_j & \beta_j \\ -\sigma_j & \delta_j \end{pmatrix} \begin{pmatrix} Q_j \\ H \end{pmatrix}, \quad \begin{pmatrix} Q_j \\ H \end{pmatrix} = \begin{pmatrix} \delta_j & -\beta_j \\ \sigma_j & a_j \end{pmatrix} \begin{pmatrix} M_j \\ L_j \end{pmatrix}$$

where

$$\sigma_j = a_1 \cdots \hat{a}_j \cdots a_n ,$$

$$\delta_j = \sum_{i \neq j} \beta_i a_1 \cdots \hat{a}_i \cdots \hat{a}_j \cdots a_n .$$

REMARK. This lemma includes the equation $H = \sigma_j M_j + a_j L_j$. This implies that H is homologous to $a_j S_j$ in $(D_2 \times S_1)_j$, which gives the confirmation, promised earlier, that our orientation of S_i is the "natural" one if a_j is positive.

Proof of Lemma. The equation

$$\sum_{i=1}^{n} \beta_i a_1 \cdots \hat{a}_i \cdots a_n = 1$$

can be rewritten $\beta_j \sigma_j + a_j \delta_j = 1$, that is

$$\det \begin{pmatrix} a_j & \beta_j \\ -\sigma_j & \delta_j \end{pmatrix} = 1 .$$

Thus the two matrix equations to be proven are mutually inverse, and we need only prove the first.

The equation $M_j = a_j Q_j + \beta_j H$ is our orientation convention for S_j. Denote for the moment $L_j' = -\sigma_j Q_j + \delta_j H$. Then $\{Q_j, H\}$ and $\{M_j, L_j'\}$ are compatibly oriented bases of $H_1(\partial(D^2 \times S^1)_j)$, since the change of basis matrix has determinant 1. But we oriented Q_j so that $Q_j \cdot H = 1$ (intersection number in $\partial(D^2 \times S_1)_j$). Thus $M_j \cdot L_j' = 1$, so L_j' is a correctly oriented longitude of S_j. To see that L_j' is the standard longitude L_j, note that since $Q_1 \cup \cdots \cup Q_n$ bounds the section $R \subset \Sigma_0$, we have the following computation in homology of Σ_0 (we take $j=1$ for simpler notation):

$$Q_1 = -(Q_2+\cdots+Q_n), \text{ so}$$

$$L_1' = -\sigma_1 Q_1 + \delta_1 H = \sigma_1(Q_2+\cdots+Q_n) + \delta_1 H$$

$$= \sum_{i=2}^{n} \alpha_2 \cdots \hat{\alpha}_i \cdots \alpha_n(\alpha_i Q_i + \beta_i H)$$

$$= \sum_{i=2}^{n} \alpha_2 \cdots \hat{\alpha}_i \cdots \alpha_n M_i .$$

Thus, L_1' is nullhomologous in $\Sigma - (D^2 \times S^1)_1$, which shows that it is the standard longitude L_1.

The last equation also implies $\ell(L_1, S_j) = \alpha_2 \cdots \hat{\alpha}_j \cdots \alpha_n$ for $j \neq 1$, since $\ell(M_i, S_j) = \delta_{ij}$. But $\ell(S_1, S_j) = \ell(L_1, S_j)$ for $j \neq 1$, so Proposition 7.4 also follows.

Our second proof of 7.4 illustrates the use of our analytic description of $(\Sigma(\alpha_1, \cdots, \alpha_n), S_1 \cup \cdots \cup S_n)$. Assume $\alpha_i \geq 1$ for $i = 1, \cdots, n$, and consider

$$\Sigma = \Sigma(\alpha_1, \cdots, \alpha_n) = \{Z \in \mathbf{C}^n | a_{i1} Z_1^{\alpha_1} + \cdots + a_{in} Z_n^{\alpha_n} = 0 ,$$

$$i = 1, \cdots, n-2\} \cap S^{2n-1} ,$$

with the S^1-action $t(Z_1, \cdots, Z_n) = (t^{q_1} Z_1, \cdots, t^{q_n} Z_n)$, where $q_i = \alpha_1 \cdots \alpha_n / \alpha_i$. Recall S_i is the intersection of Σ with the hyperplane $Z_i = 0$.

Consider the map

$$\Lambda_i : \Sigma - S_i \to S^1 , \quad \Lambda_i(Z) = Z_i / \|Z_i\| .$$

This Λ_i is S^1-equivariant for our given S^1-action on Σ and the non-effective S^1-action $t \cdot w = t^{q_k} w$ of S^1 on S^1. It follows that Λ_i is a fibration (in fact, it is the Milnor fibration at the origin for the hypersurface $\{Z_i = 0\}$ in the complex surface $V_A(\alpha_1, \cdots, \alpha_n)$). The knot (Σ, S_i)

is thus a fibered knot, with fibration Λ_i, and the fibers of Λ_i are Seifert surfaces for S_i in Σ. We can thus compute $\ell(S_i, S_j)$ as the intersection number of any such fiber with S_j, or equivalently, as the degree of $\Lambda_i|S_j : S_j \to S^1$. But S_j is an S^1-orbit of Σ with isotropy $\mathbf{Z}/(a_j)$ and we have already observed that Λ_i, and hence $\Lambda_i|S_j$, is S^1-equivariant if we use the S^1-action on S^1 with isotropy $\mathbf{Z}/q_i$. Thus $\Lambda_i|S_j$ has degree $q_i/a_j = a_1 \cdots a_n/a_i a_j$, as desired.

The same argument applies, using our analytic description, also when $(a_1, \cdots, a_n) = (0, 1, \cdots, 1)$, if one makes the obvious necessary changes of notation, and the general case of 7.4 follows from the case $a_i \geq 0$ using part 3 of 7.3.

8. *Classification of graph links and graph multilinks*

We shall represent graph links by certain diagrams, which we call "splice diagrams", of which a typical example is illustrated below.

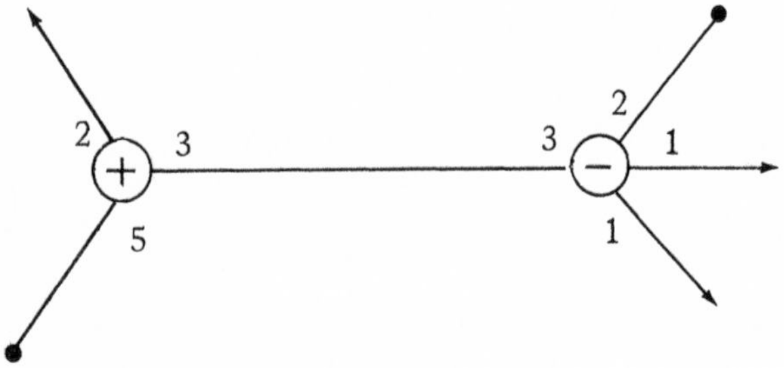

The parts out of which such a splice diagram is constructed are:

NODES:

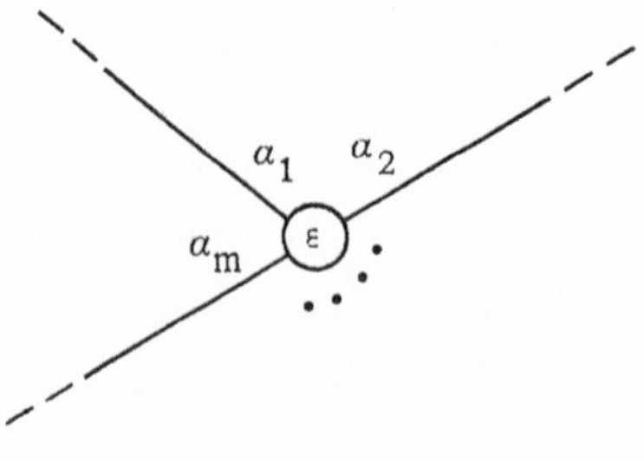

Here $m \geq 3$, the α_i are pairwise coprime integers, and ε is a sign ($+$ or $-$). (When we speak of multiplying by a vertex weight, we think of $+$ as 1 and $-$ as -1.) A node of the diagram corresponds to a Seifert manifold embedded in the link exterior; the edges incident to the node correspond to boundary components of the Seifert manifold, or, for edges that lead to boundary vertices or arrowhead vertices, to boundaries of tubular neighborhoods of fibers.

Boundary vertices: ———●

A boundary vertex corresponds to a solid torus, the tubular neighborhood of some fiber in a Seifert manifold as above.

Arrowhead vertices: ———→

These will correspond to link components of the link being represented.

When we speak of "the vertices of the diagram", we will include nodes as well as boundary vertices and arrowhead vertices.

SEIFERT LINKS:

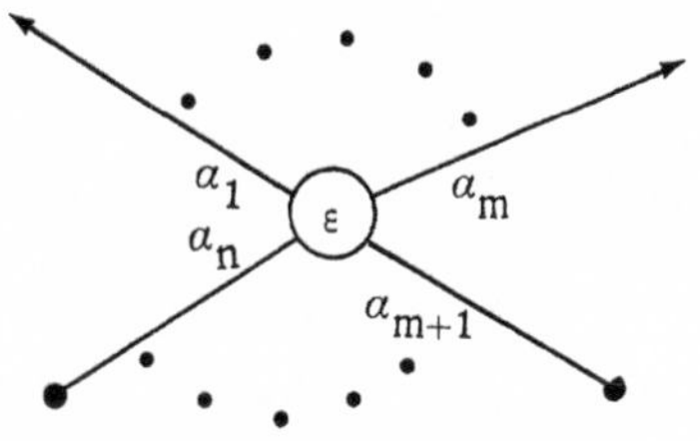

represents the Seifert link

$$(\varepsilon\, \Sigma(\alpha_1, \cdots, \alpha_n), S_1 \cup \cdots \cup S_m)$$

INDUCTIVE STEP (SPLICING): If we have already represented $L^{(1)} = (\Sigma^{(1)}, K^{(1)})$ and $L^{(2)} = (\Sigma^{(2)}, K^{(2)})$ by diagrams

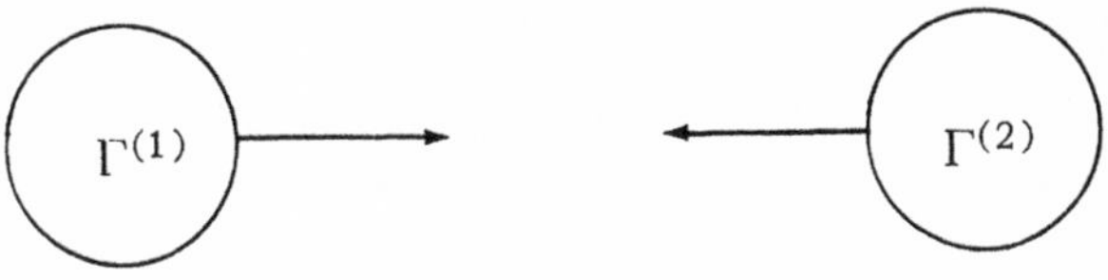

where the vertices shown represent link components $S^{(1)} \subset K^{(1)}$ and $S^{(2)} \subset K^{(2)}$, then

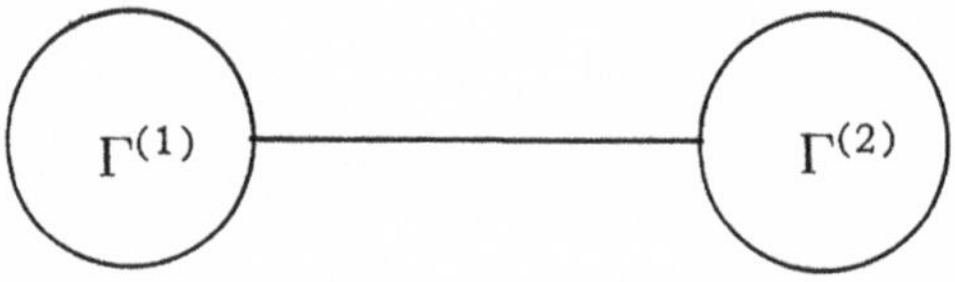

represents $L^{(1)} \underset{s^{(1)}\ s^{(2)}}{\underline{\qquad\qquad}} L^{(2)}$.

SPECIAL CASES: The following three splice diagrams represent the indicated links in S^3. Note that our notation here is consistent with the "inductive step" above.

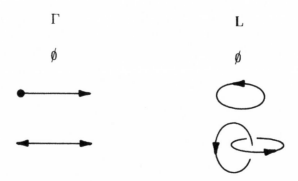

Γ	L
$\emptyset$	$\emptyset$

DISJOINT SUM: Disjoint union of splice diagrams represents disjoint sum of links.

In view of Theorem 7.3, parts 1) and 3), every Seifert link can be represented by a splice diagram. Hence, clearly *any* graph link can be represented by a splice diagram. However, it is also clear from Theorem 7.3 that this splice diagram is not unique, even for Seifert links. We shall describe this nonuniqueness precisely in Theorem 8.1, but first we extend our notation to cover graph multilinks.

If Γ is a splice diagram representing the graph link $L = (\Sigma, S_1 \cup \cdots \cup S_n)$, denote by $\Gamma(\underline{m}) = \Gamma(m_1, \cdots, m_n)$ the same diagram with a multiplicity m_i assigned to the i-th arrowhead and let it represent the multilink $L(\underline{m}) = (\Sigma, m_1 S_1 \cup \cdots \cup m_n S_n)$. We write $L = L(\Gamma)$ and $L(\underline{m}) = L(\Gamma(\underline{m}))$. We shall obtain the classification of graph links as the special case where each m_i equals ± 1, of the classification of graph multilinks by splice diagrams.

We say two splice diagrams $\Gamma(\underline{m})$ and $\Gamma'(\underline{m}')$ are equivalent, written $\Gamma(\underline{m}) \approx \Gamma'(\underline{m}')$, if $L(\Gamma(\underline{m})) \cong L(\Gamma'(\underline{m}'))$. In the following theorem $\Gamma_i(\underline{m}_i)$ may represent a piece of a splice diagram rather than a complete splice diagram.

THEOREM 8.1.

1). $\Gamma(\underline{m}) \approx \Gamma(-\underline{m})$; *in fact, graph links are invertible.*

2).
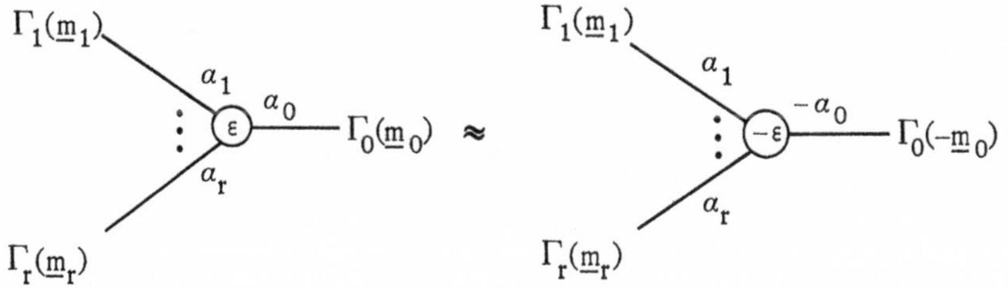

3).
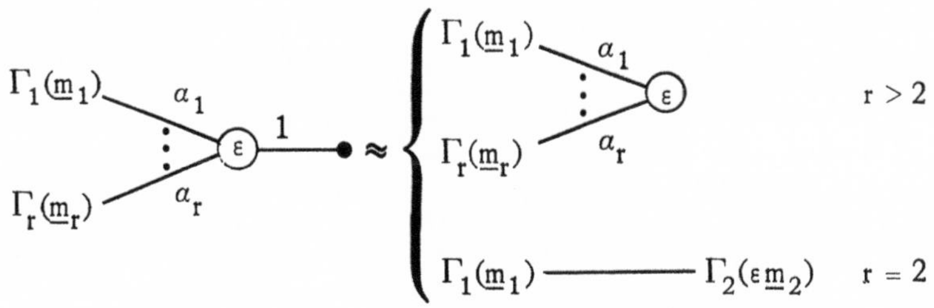

4).
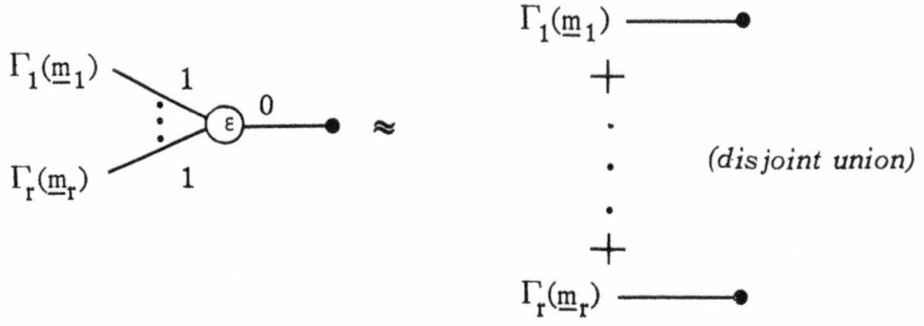

5). $\Gamma(\underline{m})$ +  $\approx \Gamma(\underline{m})$
(*disjoint union*)

6). *If* $\alpha_0 \alpha_0' = \varepsilon\varepsilon' \alpha_1 \cdots \alpha_r \alpha_1' \cdots \alpha_s'$ *then*

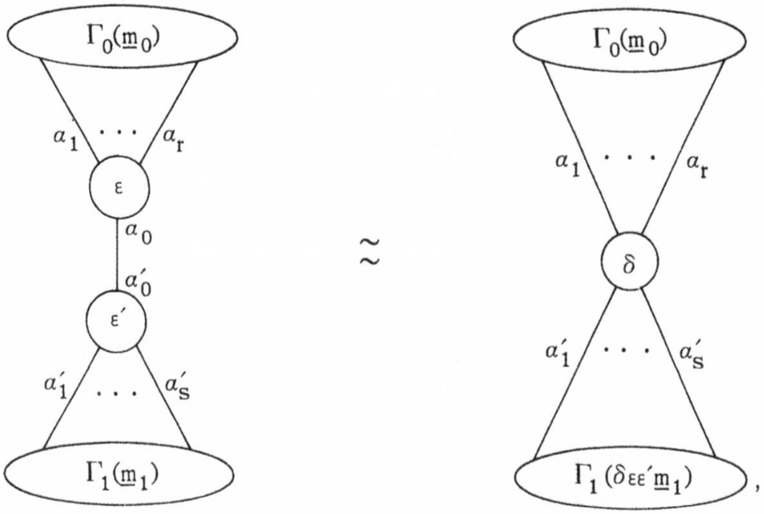

where $\delta = \pm 1$ *is chosen so that*

$$\alpha_0 = \delta\varepsilon\alpha_1' \cdots \alpha_s' \,,$$

$$\alpha_0' = \delta\varepsilon'\alpha_1 \cdots \alpha_r \,.$$

(*Note that* δ *exists, since* α_0 *is prime to* $\alpha_1 \cdots \alpha_r$ *and* α_0' *is prime to* $\alpha_1' \cdots \alpha_s'$.)

Let us say that a splice diagram is *minimal* if no equivalent diagram has fewer edges.

THEOREM 8.2 (Classification). *Up to changes of sign, as in parts 1, 2 of Theorem 8.1, there is a unique minimal splice diagram in each equivalence class. If* $\Gamma(\underline{m})$ *is a splice diagram not containing any of the configurations on the left-hand sides of parts 3-6 of Theorem 8.1, then* $\Gamma(\underline{m})$ *is minimal.*

Thus to check whether given splice diagrams are equivalent, one reduces them to minimal diagrams by applying Theorem 8.1, parts 3-6, always in the "right direction"; and then checks to see whether they differ only by a change of sign as in parts 1 and 2. In particular, we see from Theorem 8.2 that any two equivalent splice diagrams are related by a series of transformations as in 8.1, and their inverses.

If one wishes, one can reduce the classification to a *normal form* for splice diagrams:

COROLLARY 8.3. *Any irreducible graph multilink can be represented by a splice diagram* $\Gamma(\underline{m})$ *satisfying the following conditions; such a* $\Gamma(\underline{m})$ *is unique up to replacing it by* $\Gamma(-\underline{m})$.

a). $\Gamma(\underline{m})$ *is minimal;*

b). *All edge weights* α_i *are non-negative;*

c). *If an edge weight is zero, the adjacent node has weight* $+$.

The proof of the corollary is clear. In the case of an empty link this corollary is precisely the statement of Siebenmann's classification of irreducible plumbed homology spheres [Si].

Proof of 8.1. For part 1) note that we already know that Seifert links are invertible (Proposition 7.3, part 2). But the result of splicing invertible links is again invertible, since an inverting homeomorphism for a link must take longitudes to longitudes and meridians to meridians, reversing orientations of both, so it is compatible with splicing. In fact, we see that graph links are strongly invertible: the inverting homeomorphism can be chosen to be an involution.

Parts 2) and 3) follow directly from parts 3) and 4) of Proposition 7.3, while 4) follows from part 6) of 6.3 plus part 3) of 1.1. Part 5) is trivial, so only part 6) needs discussion. Using part 2), it is easy to reduce the general case of 6) to the case $\varepsilon = \varepsilon' = +1$. Moreover, since any other splicings can be done later, we only need the following lemma.

LEMMA 8.4. *The splice diagram* $\Gamma(\underline{m})$ *below represents a Seifert link if and only if*

$$a_0 = \delta a_1' \cdots a_s' ,$$

$$a_0' = \delta a_1 \cdots a_r$$

for some $\delta = \pm$. *In this case* $\Gamma(\underline{m}) \approx \Gamma'(\underline{m}')$.

$\Gamma(\underline{m}) =$

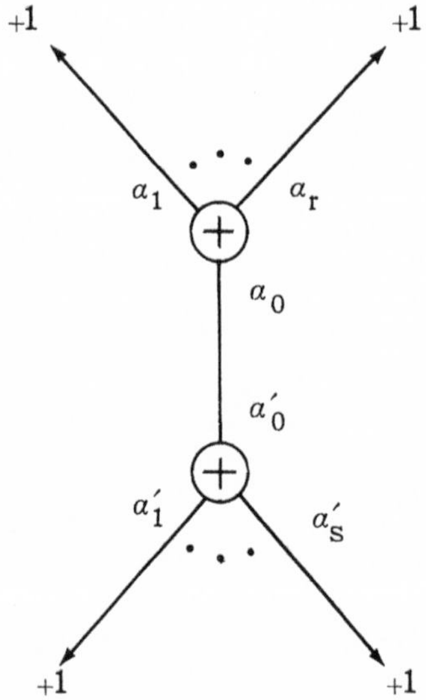

$\Gamma'(\underline{m}') =$

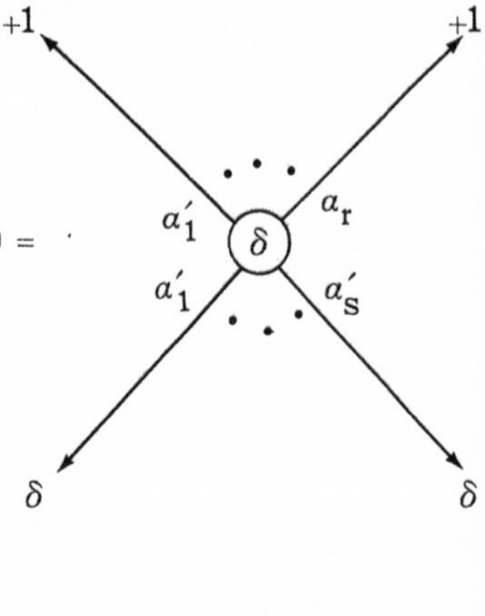

Proof. $L(\Gamma(\underline{m}))$ is the result of splicing the Seifert link

$$(\Sigma, K) = (\Sigma(a_0, \cdots, a_r), S_0 \cdots S_r)$$

to

$$(\Sigma, K') = (\Sigma(a_0', \cdots, a_s'), S_0' \cdots S_s')$$

along S_0 and S_0'. Let M_0, L_0, H_0 be as in Lemma 7.5, so

$$H_0 = a_1 \cdots a_r M_0 + a_0 L_0 .$$

Let M_0', L_0', H_0' be analogously chosen for (Σ, K'), so

$$H_0' = a_1' \cdots a_s' M_0' + a_0' L_0' .$$

After splicing, M_0 is identified with L_0' and L_0 with M_0'. Thus, the splicing can be made to match up the Seifert fibrations of the link exteriors $\Sigma - \operatorname{int} N(k)$ and $\Sigma' - \operatorname{int} N(K')$ if and only if

$$a_1 \cdots a_r M_0 + a_0 L_0 = \pm(a_1' \cdots a_s' L_0' + a_0' M_0')$$

in the homology of the torus T^2 along which we splice. Thus, if this equation is not satisfied, then the splice decomposition is the minimal splice decomposition of $L(\Gamma(\underline{m}))$, so $L(\Gamma(\underline{m}))$ cannot be a Seifert link. On the other hand, if the equation is satisfied, then certainly $L(\Gamma(\underline{m}))$ and $L(\Gamma'(\underline{m}'))$ are the same up to orientations of total space and link components. That orientations agree is seen by comparing linking numbers using Propositions 7.4 and 1.2 (see also section 10).

Theorem 8.2 now follows easily from Theorem 2.2 since the above argument shows that a splice diagram not containing any of the configurations on the left-hand sides of the equivalences 3-6 of Theorem 8.2 represents the minimal decomposition of the graph link, reducing 8.2 to the classification of Seifert links, done in section 7.

9. *Solvable and algebraic links as graph links in* S^3

PROPOSITION 9.1. *If*

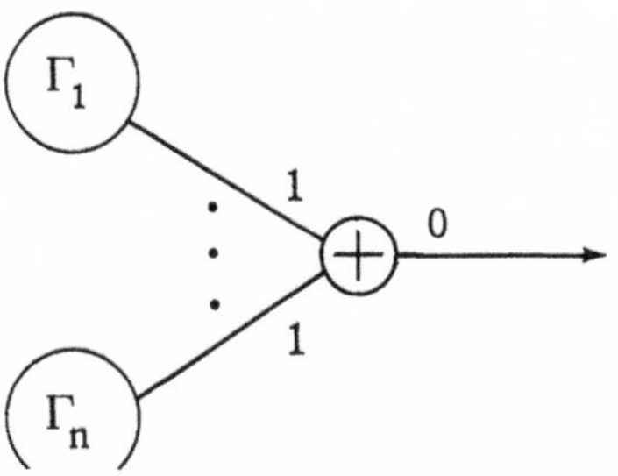

and

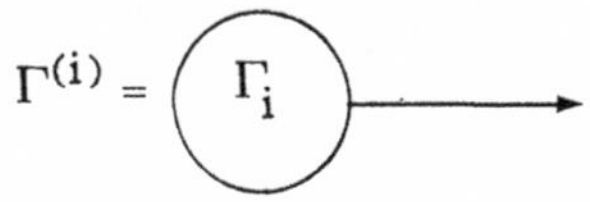

then $\mathbf{L}(\Gamma) = \mathop{\#}\limits_{i-1}^{n} \mathbf{L}(\Gamma^{(i)})$, *summed along the link components corresponding to the vertices of the* $\Gamma^{(i)}$ *shown.*

If

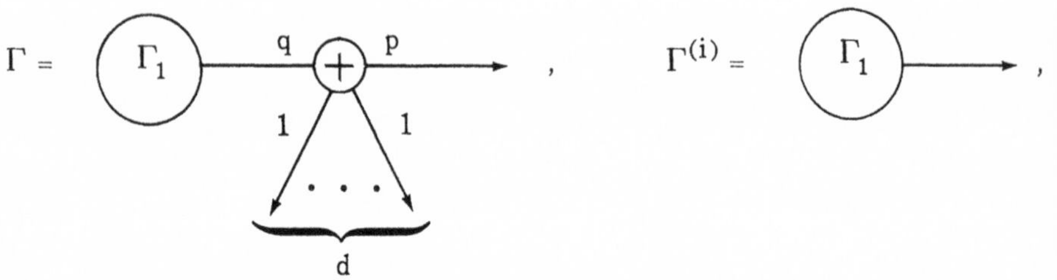

then $\mathbf{L}(\Gamma) = (\Sigma, K \cup d\, S(p,q))$, *where* $(\Sigma, K) = \mathbf{L}(\Gamma^{(1)})$ *and* $S \subset K$ *is the component corresponding to the vertex of* $\Gamma^{(1)}$ *shown.*

This is by Propositions 1.1 and 7.3. The second part of Proposition 9.1 implies that a diagram of the form

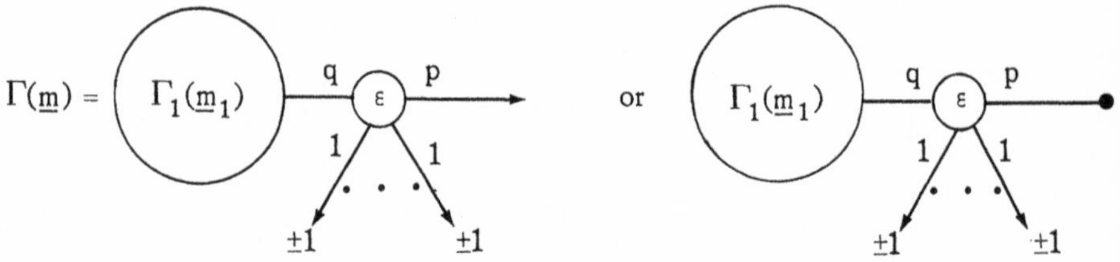

with all multiplicities ± 1 always represents a link which is the result of a cabling operation on a simpler link, with subsequent reversal of orientations of some components if necessary.

THEOREM 9.2. *If* $\mathbf{L} = (S^3, K)$ *is a graph link, then it is a solvable link, that is, it can be built up from trivial knots in* S^3 *using only cabling and summing operations.*

Proof. We may assume L is irreducible. We can also disregard orientations of link components, so we shall work with splice diagrams having no multiplicity weights.

Let Γ be the normal form splice diagram for L, as in Corollary 8.3. We argue by induction on the number of nodes in Γ. If we call a node of the form

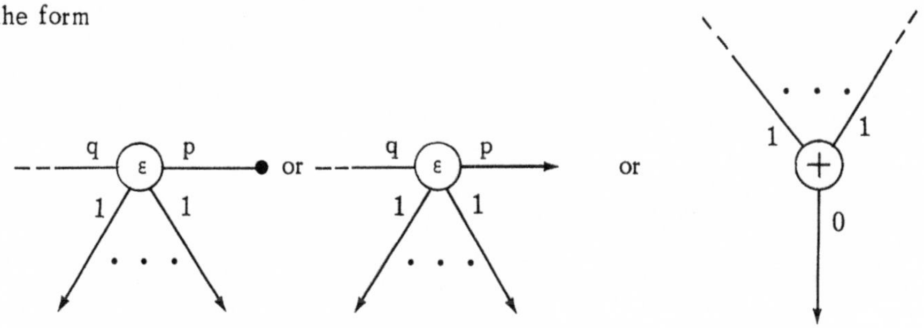

in Γ a cable node or a sum node for the purpose of this proof, then it suffices to show Γ has such a node.

Let Γ' be Γ with all arrowheads replaced by boundary vertices. Then $L(\Gamma') = (S^3, \phi)$, so Γ' reduces to the empty diagram by parts 3) to 6) of Theorem 8.1. In fact, part 6) never occurs by minimality of Γ. Let v be the first node of Γ which vanishes during some process of reducing Γ'. If v vanishes by part 4) of 8.1, it is a sum node. Otherwise, it vanishes as in part 3) of 8.1 with $r = 2$, so v is as in the following picture.

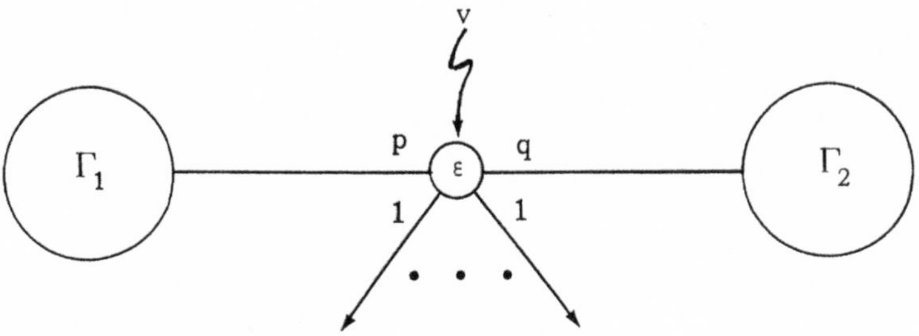

If either Γ_1 or Γ_2 has only one vertex, then v is a cable node. Otherwise, the diagram

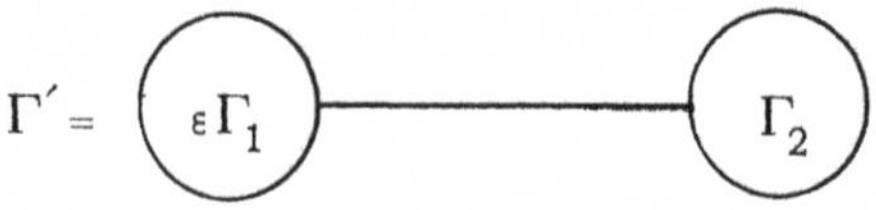

has a cable node or sum node by induction hypothesis, and this node will be a cable node or sum node for Γ.

As an immediate consequence, we see that the minimal splice diagram for a solvable *knot* has a rather simple form:

COROLLARY 9.3. *If*

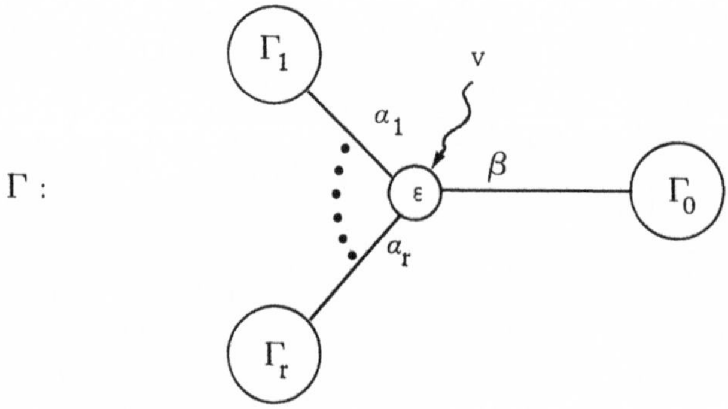

is a minimal splice diagram for a solvable knot, and the one arrowhead vertex is in Γ_0, *then either* $\beta = 0$, $\alpha_1 = \cdots = \alpha_r = 1$ *(and the node represents a connected sum operation) or* $r = 2$ *and (exchanging* Γ_1 *and* Γ_2 *if necessary)* Γ *has the form*

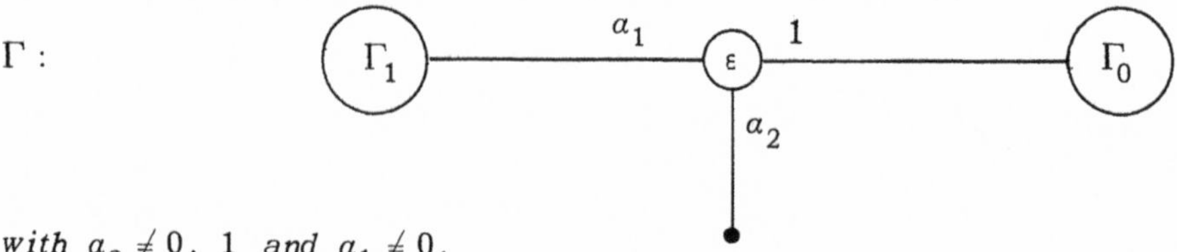

with $\alpha_2 \neq 0, 1$ *and* $\alpha_1 \neq 0$.

A solvable link may be constructible via cabling and summing in many different ways, but this will always be evident from the splice diagram Γ. For instance consider

$\Gamma =$

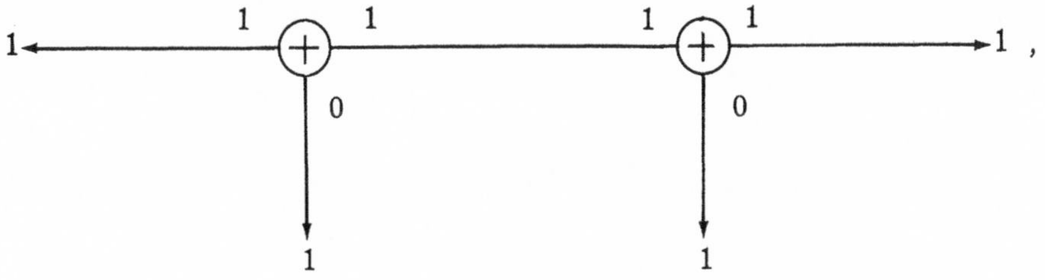

for which

$L(\Gamma) =$ 

The following gives several ways of building this link, and on the right a suggested notation for keeping track of this in Γ, by "cuts" to show where summing occurred and arrows on edges to show the direction of building $L(\Gamma)$ by cabling:

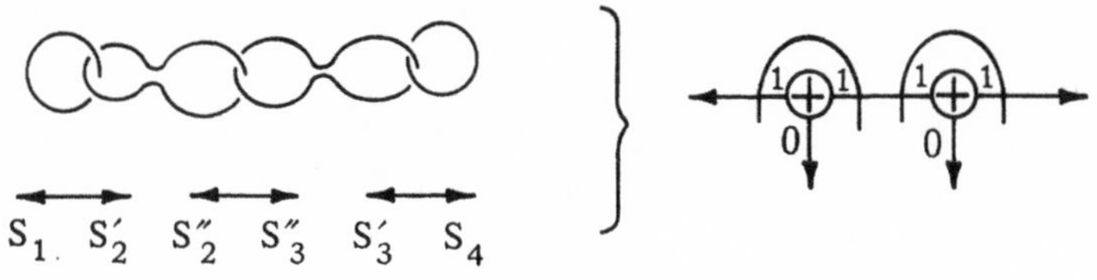

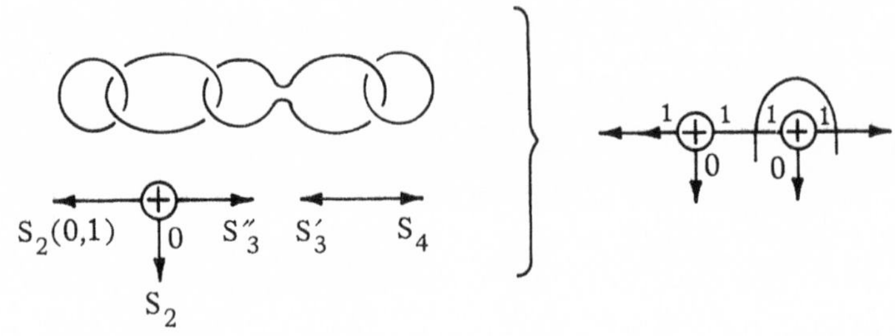

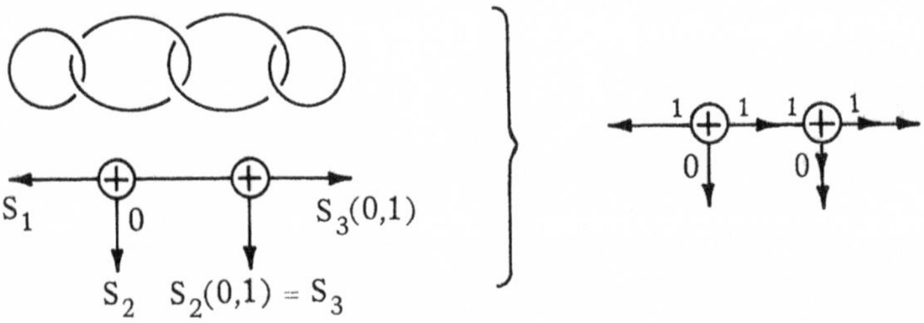

Here is a more complicated example with its normal form splice diagram.

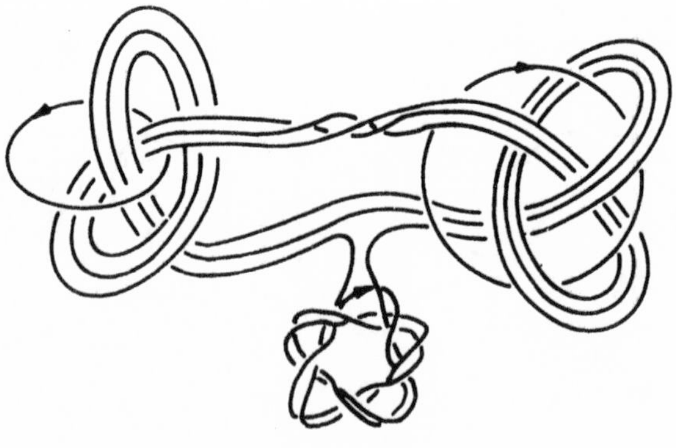

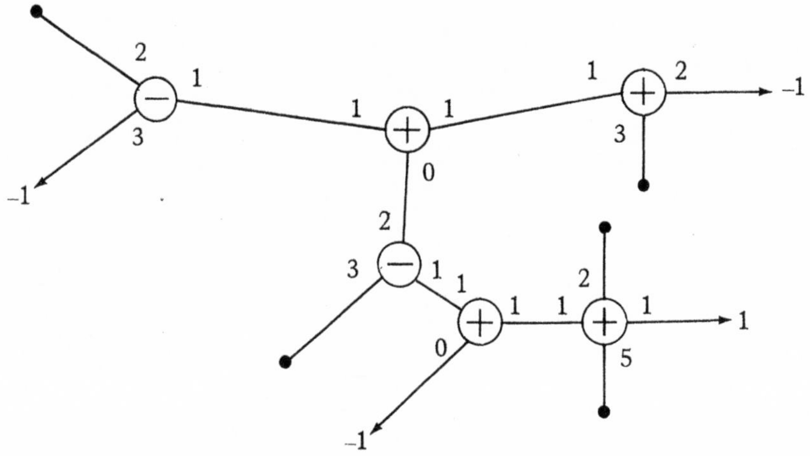

It can be built up as follows:

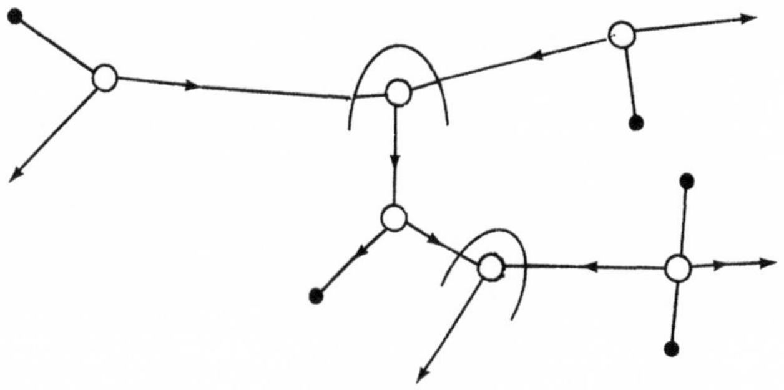

The algebraic case

For a solvable link to be an algebraic link, certain inequalities must be satisfied by the weights in the graph.

THEOREM 9.4. *The solvable link* $L(\Gamma)$ *is an algebraic link if*

i) *all weights are positive (both edge weights and vertex weights)*

ii) *in a portion of* Γ *as follows,*

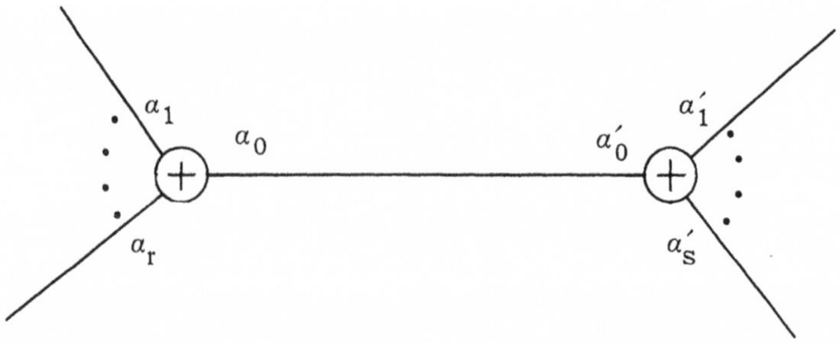

one has $\alpha_0\alpha'_0 - \alpha_1 \cdots \alpha_r \alpha'_1 \cdots \alpha'_s > 0$. *Conversely, if* L *is an algebraic link then its normal form graph satisfies the above conditions.*

Proof. This is a direct translation of the standard inequalities for Puiseux data. These inequalities come from the fact that if t is small and $a > b$ then $t^a < t^b$.

Generalizing, if one defines an *algebraic graph link* as the link of a pair (V, C) consisting of an algebraic curve C in a normal algebraic surface V at a point $p \in C$ which possibly singular for both, then conditions i) and ii) above characterize algebraic graph links among all graph links. This will be proved in section 24.

Chapter III

INVARIANTS

10. *Linking numbers for graph links and multilinks*

Let Γ be a splice diagram with no multiplicity weights, representing a graph link $L(\Gamma) = (\Sigma, S_1 \cup \cdots \cup S_n)$ and let $\Gamma(\underline{m}) = \Gamma(m_1, \cdots, m_n)$ be the splice diagram for $(\Sigma, m_1 S_1 \cup \cdots \cup m_n S_n)$. We fix this notation throughout this section.

Moreover, we shall number the vertices of Γ as

$$v_1, \cdots, v_n,\ v_{n+1}, \cdots, v_k,$$

with $v_1, \cdots, v_n$ being the arrowheads and $v_{n+1}, \cdots, v_k$ being the remaining vertices. For $i = n+1, \cdots, k$, let S_i be a nonsingular fiber of the Seifert component corresponding to v_i, if v_i is a node, and let S_i be the (possibly) singular fiber corresponding to v_i if v_i is a boundary vertex. We always orient S_i as in section 7. Thus, $S_{n+1}, \cdots, S_k$ can be thought of as "virtual components" of $L(\Gamma)$: they are not really components, but are nevertheless determined up to isotopy by our description of the link in terms of Γ, and are thus intrinsic to $L(\Gamma)$, up to isotopy and orientation, if Γ is a minimal splice diagram.

For any two distinct vertices v_i and v_j of Γ, let σ_{ij} be the simple path in Γ joining v_i to v_j, including v_i and v_j, and define

$$\ell_{ij} = (\text{product of all signs of nodes on } \sigma_{ij}) \cdot (\text{product of all edge weights adjacent to these nodes but not on } \sigma_{ij})$$

as in the following examples (the path σ_{ij} is drawn bold):

$$\ell_{ij} = (+1 \cdot -1) \cdot (1 \cdot 3 \cdot 2 \cdot 3)$$
$$= -18$$

$$\ell_{ij} = (-1 \cdot +1 \cdot +1 \cdot -1) \cdot (2 \cdot 3 \cdot 1 \cdot 2 \cdot 5)$$
$$= 60$$

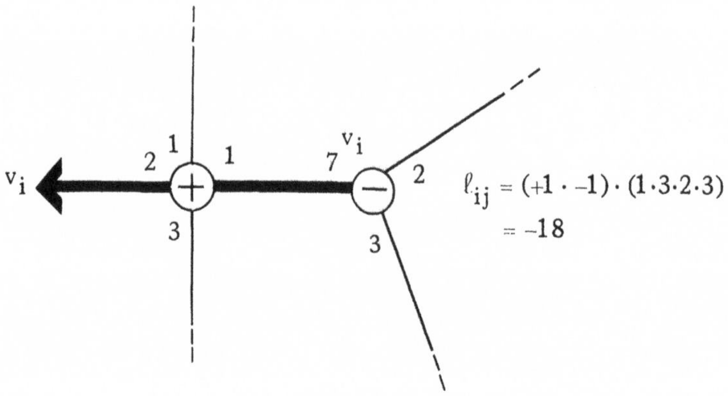

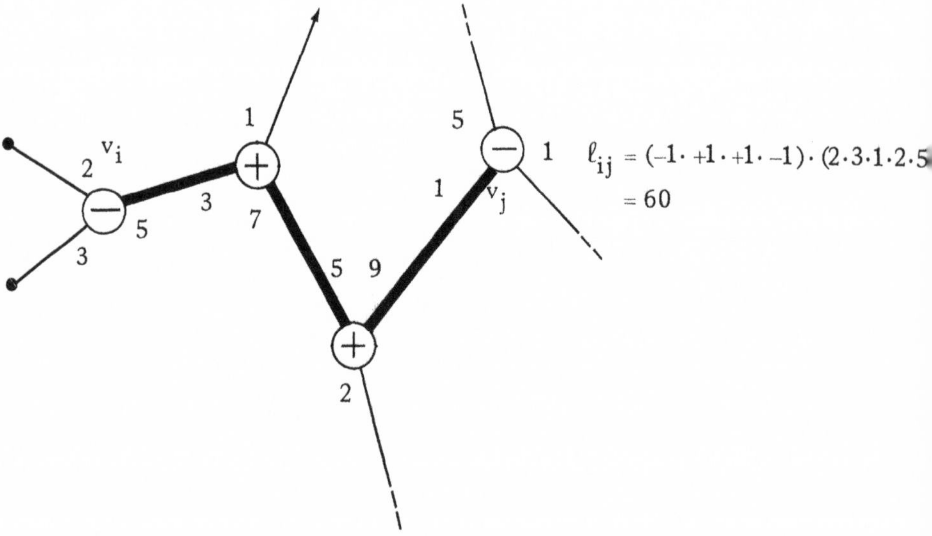

Of course, if Γ is disconnected, σ_{ij} may not exist, in which case we put $\ell_{ij} = 0$.

THEOREM 10.1. $\ell_{ij} = \ell(S_i, S_j)$ *for* $1 \leq i < j \leq k$. *Hence*

$$\underline{m}(S_j) = m_1 \ell_{1j} + \cdots + m_n \ell_{nj} \quad for \;\; k+1 \leq j \leq n \;.$$

Proof. For the splice diagrams of Seifert links this is a restatement of Proposition 7.4. In general it thus follows by induction using Proposition 1.2.

This computation puts strong constraints on the possible mutual linking numbers of components in a graph link. We leave the proofs of the following remarks to the reader.

10.2. *Three integers* a_{12}, a_{23}, a_{13} *can occur as* $a_{ij} = \ell(S_i, S_j)$ *for some graph link* $(S^3, S_1 \cup S_2 \cup S_3)$ (i.e. *solvable link*) *if and only if one of the* a_{ij} *divides the product of the other two.*

10.3. *Six integers* a_{ij} $1 \le i < j \le 4$ *can occur as* $a_{ij} = \ell(S_i, S_j)$ *for some graph link* $(\Sigma, S_1 \cup S_2 \cup S_3 \cup S_4)$ *if and only if, after some permutation of the indices* $i = 1,2,3,4$, *it is true that* $a_{12}a_{34} = a_{13}a_{24}$

For a five component link there will be the constraint of 10.3) on each 4-component sublink plus additional constraints. For example:

10.4. *Given* a_{ij}, $1 \le i < j \le 5$, *with* $a_{ij} \ne 0$ *if and only if* $(i-j) \equiv \pm 1$ (mod 5), *there is no realization by a graph link.*

In (10.4) all four-component constraints are satisfied.

Linking number constraints are often a fast way of showing that a given link is *not* a graph link. For example, the two component link in S^3

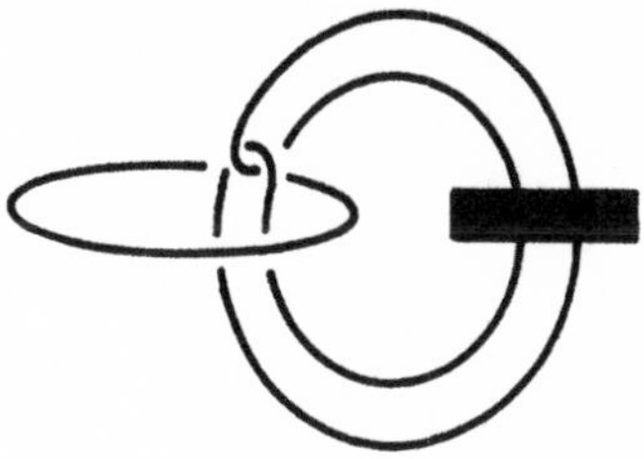

where the box represents anything, is not solvable, since its exterior can be covered by the exterior of

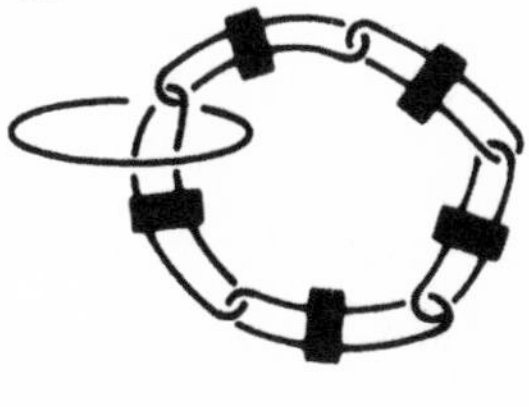

which has a sublink to which (10.4) applies. This example is not hard to see in other ways, but it does stress the way in which very coarse linking information can force the existence of a simple splice component in the minimal splice decomposition of a link.

Linking numbers can be computed similarly for other curves in Σ. Suppose v_p and v_q are adjacent vertices of Γ, connected by an edge E, as in the following picture.

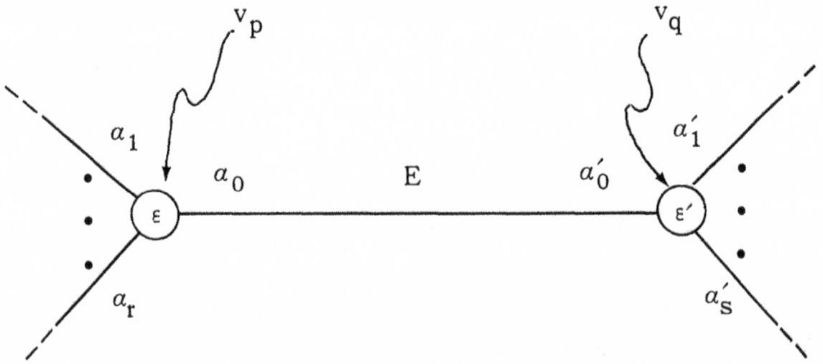

E corresponds to a separating torus T in the link exterior Σ_0. This situation corresponds to a splice decomposition

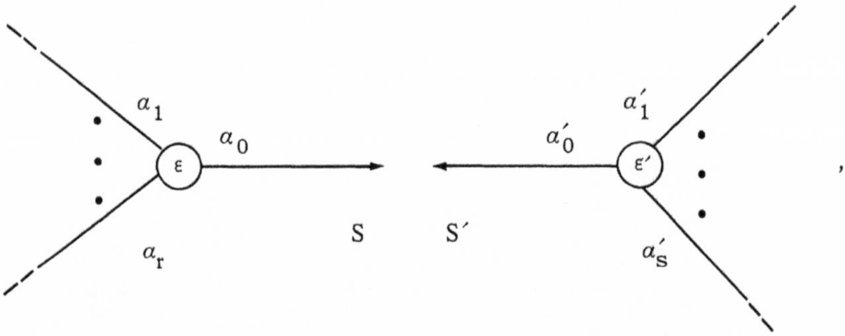

and in T we have curves $M = L'$ and $L = M'$ which are respectively meridian and longitude of S (or longitude and meridian of S′).

For any vertex v_i of Γ, let σ_{Ei} be the path in Γ joining edge E to vertex v_i, including E and v_i, and let

ℓ_{Ei} = (product of all signs of nodes on σ_{Ei})$\cdot$(product of all edge weights adjacent to these nodes but not on σ_{Ei}).

THEOREM 10.5. *According as vertex* v_i *occurs to the left or right of edge* E *in* Γ, *we have*

$$\begin{Bmatrix} \ell(L, S_i) = \ell_{Ei} \\ \\ \ell(M, S_i) = 0 \end{Bmatrix} \quad \text{or} \quad \begin{Bmatrix} \ell(L, S_i) = 0 \\ \\ \ell(M, S_i) = \ell_{Ei} \end{Bmatrix}$$

Let $v_1, \cdots, v_t$ and $v_{t+1}, \cdots, v_n$ be respectively the arrowheads of Γ to the left and to the right of E.

COROLLARY 10.6. *Let* $m' = m_1 \ell_{E,1} + \cdots + m_t \ell_{E,t}$ *and* $m = m_{t+1} \ell_{E,t+1} + \cdots + m_n \ell_{E,n}$. *Then the multilink splice summands of* $L(\Gamma(\underline{m}))$ *resulting from splitting along* T *are given by the graphs*

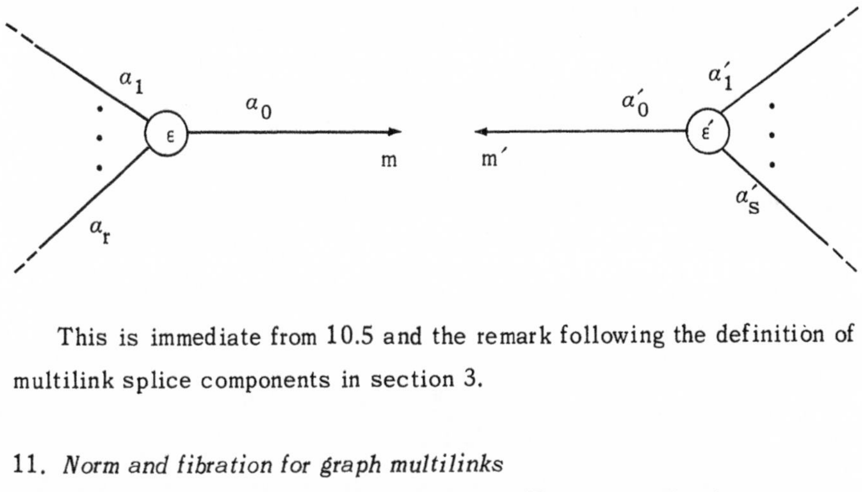

This is immediate from 10.5 and the remark following the definition of multilink splice components in section 3.

11. *Norm and fibration for graph multilinks*

We retain the notation of section 10, so Γ has arrowheads $v_1, \cdots, v_n$ and remaining vertices $v_{n+1}, \cdots, v_k$, and the non-arrowhead vertices correspond to "virtual components" $S_{n+1}, \cdots, S_k$ in Σ. Let δ_j denote the degree of the j-th vertex in Γ (number of incident edges, so $\delta_j = 1$ for a boundary vertex or arrowhead, $\delta_j \geq 3$ for a node). We consider a multilink $L(\Gamma(\underline{m})) = L(\Gamma(m_1, \cdots, m_n))$ on $L(\Gamma)$.

THEOREM 11.1. *Assume* $L(\Gamma)$ *is irreducible, and not the unknot* (S^3, S^1). *Then the norm of* $L(\Gamma(m)) = L(\Gamma(m_1, \cdots, m_n))$ *is*

$$\|\underline{m}\| = \sum_{j=n+1}^{k} (\delta_j - 2)\,|\underline{m}(S_j)| = \sum_{j=n+1}^{k} (\delta_j - 2)\,|m_1 \ell_{1j} + \cdots + m_n \ell_{nj}|$$

THEOREM 11.2. *Assume* $L(\Gamma)$ *is irreducible and* Γ *is a minimal splice diagram for* $L(\Gamma)$, *and* $\Gamma \neq \emptyset$ or $\leftrightarrow$. *Then* $L(\Gamma(m)) = L(\Gamma(m_1, \cdots, m_n))$ *is a fibered multilink if and only if* $\underline{m}(S_j) = m_1 \ell_{1j} + \cdots + m_n \ell_{nj}$ *is nonzero for* $j = n+1, \cdots, k$. *This condition for fiberability is sufficient even if* Γ *is not minimal.*

This theorem implies that if the set

$$\{\underline{m} \in H^1(\Sigma_0; \mathbf{Z}) \mid L(\Gamma(\underline{m})) \text{ is fibered}\}$$

is nonempty, then it is the complement of a finite collection of hyperplanes in $H^1(\Sigma_0; \mathbf{Z})$. An analogous statement holds, using the same proof, for any graph manifold, and the following theorem can be similarly generalized.

THEOREM 11.3. *If* $L(\Gamma(\underline{m}))$ *is fibered and* d *is the largest integer which divides* $\underline{m}$ (*that is,* $d = \gcd(m_1, \cdots, m_n)$), *then the fiber* F *has exactly* d *components which are cyclically permuted by the monodromy* $h: F \to F$. *The characteristic polynomial of* $H_1(h): H_1(F) \to H_1(F)$ *is*

$$\Delta_1(t) = (t^d - 1) \prod_{j=n+1}^{k} (t^{|\underline{m}(S_j)|} - 1)^{\delta_j - 2}$$

REMARK. In 11.3 Γ need not be assumed minimal, but then terms of the form $(t^0 - 1)^d$ may appear in the formula for $\Delta(t)$. The total exponent of such terms is zero, and the formula should be interpreted by formally cancelling these terms. If Γ is minimal, this will not arise, by Theorem 11.2.

Proofs. We shall prove all three theorems together. We may assume for all three theorems that Γ is a minimal splice diagram, since the operations of Theorem 8.1 do not alter the equations to be proved in 11.1 and 11.3 if $L(\Gamma)$ is irreducible. Moreover, by Theorems 3.3, 4.2, and 4.3, it suffices to prove the results for Seifert links.

The case $\Gamma(\underline{m}) = m_1 \leftrightarrow m_2$ is trivial, so, after reversing orientation if necessary, we may assume that

$\Gamma(\underline{m}) =$

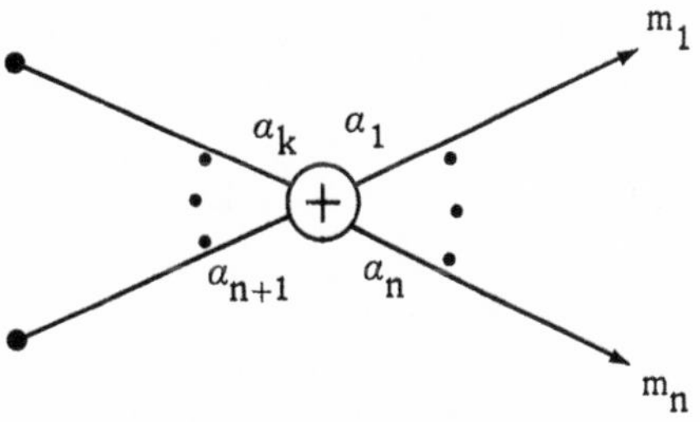

Changing notation slightly, we number the central node $k+1$ and, as in section 7, we define

$$q_i = \alpha_1 \cdots \hat{\alpha}_i \cdots \alpha_k \quad \text{for } 1 \leq i \leq k .$$

By Theorem 10.1

$$\begin{aligned} \ell_{ij} &= q_i/\alpha_j & & 1 \leq i \leq n \quad n+1 \leq j \leq k , \\ &= q_i & & 1 \leq i \leq n , \quad j = k+1 . \end{aligned}$$

Thus, if we put

$$\ell = \sum_{i=1}^{n} q_i m_i ,$$

then $\underline{m}(S_j) = \ell/\alpha_j$ for $n+1 \leq j \leq k$ and $\underline{m}(S_{k+1}) = \ell$.

The equation to be proved for 11.1 is thus

$$\|\underline{m}\| = \left(k-2 - \sum_{j=n+1}^{k} 1/a_j\right) |\ell| ,$$

while 11.2 says $L(\Gamma(\underline{m}))$ is fibered if and only if $\ell \neq 0$.

We use the analytic model $\Sigma = \Sigma(a_1, \cdots, a_k) \subset \mathbf{C}^k$ of section 7. Recall that $S_i = \{Z_i = 0\}$ in that model. Let $\Sigma' = \Sigma - (S_1 \cup \cdots \cup S_n)$. We claim that $\Lambda : \Sigma' \to S^1$, given by

$$\Lambda(Z_1, \cdots, Z_n) = Z_1^{m_1} \cdots Z_n^{m_n} / |Z_1^{m_1} \cdots Z_n^{m_n}| ,$$

is a representative for $\underline{m} \in H^1(\Sigma') = [\Sigma', S^1]$. Indeed, $\Lambda = \Lambda_1^{m_1} \cdots \Lambda_n^{m_n}$, where $\Lambda_i(Z) = Z_i/|Z_i|$, and since the group structure in $[\Sigma', S^1]$ is induced from the group structure of S^1, we need only check our claim for each Λ_i, where it follows from the fact that Z_i is an oriented transverse coordinate to S_i. In fact, in a neighborhood of a point of S_i any fiber $\Lambda^{-1}(t)$ of Λ is given locally by $\arg(Z_i^{m_i} \phi(Z)) = \text{constant}$, where ϕ is a nonzero analytic function on the neighborhood. $\Lambda^{-1}(t)$ thus has the correct local structure near each S_i to be a Seifert surface for $L(\Gamma(\underline{m}))$, so if it is nonsingular, it is a Seifert surface.

Since the S^1-action on Σ' is given by

$$t(Z_1, \cdots, Z_k) = (t^{q_1} Z_1, \cdots, t^{q_k} Z_k) ,$$

we have

$$\Lambda(t \cdot Z) = t^{q_1 m_1 + \cdots + q_n m_n} \Lambda(Z) = t^{\ell} \Lambda(Z) .$$

Assume first that $\ell \neq 0$. Then the fibers of Λ are transverse to the S^1-action on Σ', so Λ is a fibration. Moreover, a typical fiber F of Λ meets each nonsingular orbit of the S^1-action on Σ' in $|\underline{m}(S_{k+1})| = |\ell|$ points and meets the singular orbit S_j, for $n < j \leq k$, in $|\underline{m}(S_j)| = |\ell/a_j|$

points. Thus F is an $|\ell|$-fold branched cover of $F' = \Sigma'/S^1$, branched over k–n points with branch indices $\alpha_{n+1}, \cdots, \alpha_k$. Since F' is an n-fold punctured sphere, the standard formula for the Euler characteristic of a branched cover gives

$$\chi(F) = |\ell| \left(\chi(F') - \sum_{j=n+1}^{k} \frac{\alpha_j - 1}{\alpha_j} \right) = |\ell| \left(2 - n - \sum_{j=n+1}^{k} \frac{\alpha_j - 1}{\alpha_j} \right)$$

$$= |\ell| \left(2 - k + \sum_{j=n+1}^{k} 1/\alpha_j \right),$$

so

$$\|\underline{m}\| = \chi_{-}(F) = |\ell| \left(k - 2 - \sum_{j=n+1}^{k} 1/\alpha_j \right).$$

Now if $\ell = 0$ then the fibers of Λ are unions of S^1-orbits. Thus, any nonsingular fiber F of Λ is a union of annuli, whence $\chi_{-}(F) = 0$, so $\|\underline{m}\| = 0$, completing the proof of 11.1.

Moreover, 11.2 also follows, for if $L(\Gamma(\underline{m}))$ were fibered when $\ell = 0$, its fiber would have to consist of disks and/or annuli, since we have shown $\|\underline{m}\| = 0$ in this case. The link exterior, being connected, would thus be $D^2 \times S^1$ or $(I \times S^1) \times S^1$. This forces $\Gamma = \bullet\!\!\rightarrow$ or $\leftrightarrow$. The second of these diagrams was excluded in the statement of the theorem, while the theorem is trivial for the first.

Before completing the proof of 11.3 we summarize from the above proof:

LEMMA 11.4. *If* $L(\Gamma(\underline{m}))$ *is the Seifert multilink of the above proof and* $\ell \neq 0$, *then* $L(\Gamma(\underline{m}))$ *is fibered, its typical fiber* F *is an* $|\ell|$-*fold branched cyclic cover of the* n-*punctured sphere* $F' = \Sigma/S^1$ *branched over precisely* k–n *points* $x_{n+1}, \cdots, x_k \in F'$ *with branch indices* $\alpha_{n+1}, \cdots, \alpha_k$, *respectively, and the monodromy* $h: F \to F$ *of the fibration is a generator of the group* $\mathbf{Z}/|\ell|$ *of covering transformations.*

The claim about $h: F \to F$ is verified by observing that h can be described in the above proof by sliding F along the S^1-action on Σ'. In the above proof F was an open Seifert surface in the link complement, but we can equally well consider it to be a closed fiber in the link exterior.

To prove 11.3 for our Seifert multilink $L(\Gamma(\underline{m}))$ we must show

$$\Delta_* = (t^{|\ell|}-1)^{k-2} \prod_{j=n+1}^{k} (t^{|\ell/a_j|}-1)^{-1},$$

where Δ_* is the characteristic polynomial quotient of Theorem 4.3 for $L(\Gamma(\underline{m}))$.

Now for any simplicial map $h: F \to F$ of a simplicial complex F, the rational function $\Delta_* = \Delta_0^{-1}\Delta_1\Delta_2^{-1}\Delta_3 \cdots$, is the same whether we compute the characteristic polynomials Δ_i on the level of homology or on the chain level.

In our case F is a compact surface and h has finite order $|\ell|$. We can triangulate F h-equivariantly, so that we have an induced triangulation of $F' = \Sigma'/S^1$. If for any simplex σ of F' we denote by $n(\sigma)$ the number of simplices of F lying over σ and if $\varepsilon(\sigma) = (-1)^{\dim \sigma}$, then

$$\Delta_* = \prod_{\sigma} (t^{n(\sigma)}-1)^{-\varepsilon(\sigma)}.$$

Since $n(\sigma) = |\ell|$ for any simplex of $F_0 = F' - \{x_{n+1}, \cdots, x_k\}$, while $n(\sigma) = |\ell/a_j|$ if $\sigma = x_j$, this simplifies to

$$\Delta_* = (t^{|\ell|}-1)^{-\chi(F_0)} \prod_{j=n+1}^{k} (t^{|\ell/a_j|}-1)^{-1}$$

$$= (t^{|\ell|}-1)^{k-2} \prod_{j=n+1}^{k} (t^{|\ell/a_j|}-1)^{-1}$$

An interesting corollary of 11.3 is the following:

PROPOSITION 11.4. *If* $L(\Gamma(\underline{m}))$ *is a fibered graph multilink with* n *components then* 1 *is an exactly* (n–1)*-fold root of the characteristic polynomial* $\Delta_1(t)$.

Proof. In the notation of Theorem 11.3, the multiplicity of 1 as a root of $\Delta_1(t)$ is $1 + \sum_{j=n+1}^{k} (\delta_j - 2)$. This equals $1 + \sum_{j=1}^{k} (\delta_j - 2) + n$, which is (n–1) since a trivial induction shows that the sum of $\delta_j - 2$ over the vertices of any tree is –2.

Note that if $L(\underline{m})$ is *any* n-component fibered multilink (not necessarily a graph link) with monodromy $h: F \to F$, and if Σ_0 is the link exterior then we have an exact sequence

$$H_1(F) \xrightarrow{H_1(h)-1} H_1(F) \to H_1(\Sigma_0) \to Z \to 0 .$$

But $H_1(\Sigma_0) = Z^n$ by Alexander duality. Thus

$$\mathrm{Cok}(H_1(h)-1) = Z^{n-1} .$$

Thus the multiplicity of 1 as a root of $\Delta_1(t)$ is always at least (n–1), and equals (n–1) if and only if $H_1(h) \otimes R$ has no 2×2 or larger Jordan blocks at the eigenvalue 1.

COROLLARY 11.5. *A fibered graph multilink has only* 1×1 *Jordan blocks for the eigenvalue* 1 *of its algebraic monodromy.*

This is definitely not true for general multilinks. We leave the verification of the following example to the reader.

EXAMPLE 11.6. *The link in* S^3 :

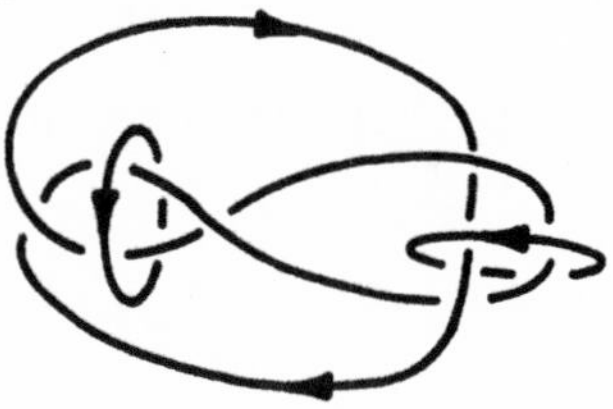

is fibered with algebraic monodromy

$$\begin{pmatrix} 1 & 1 & 0 & 0 \\ 0 & 1 & 0 & 0 \\ 0 & 0 & 1 & 1 \\ 0 & 0 & 0 & 1 \end{pmatrix}$$

In fact after deforming the picture we can draw a fiber as follows, and numbering and orienting the handles as shown gives a basis of $H_1(F)$ with the above matrix for $H_1(h)$:

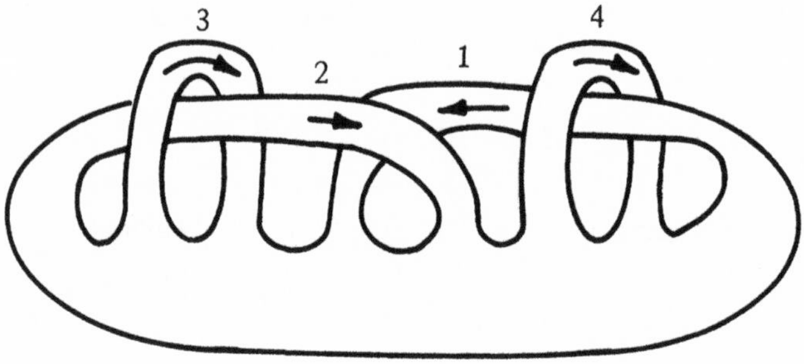

The Seifert form is

$$\begin{pmatrix} 0 & 0 & 0 & 1 \\ -1 & 0 & -1 & 0 \\ 0 & -1 & 0 & 0 \\ 1 & 0 & 0 & 0 \end{pmatrix}$$

with respect to the same basis.

Theorem 11.2 easily implies that a graph knot (graph link with only one link component) is fiberable if and only if, for every node of the form

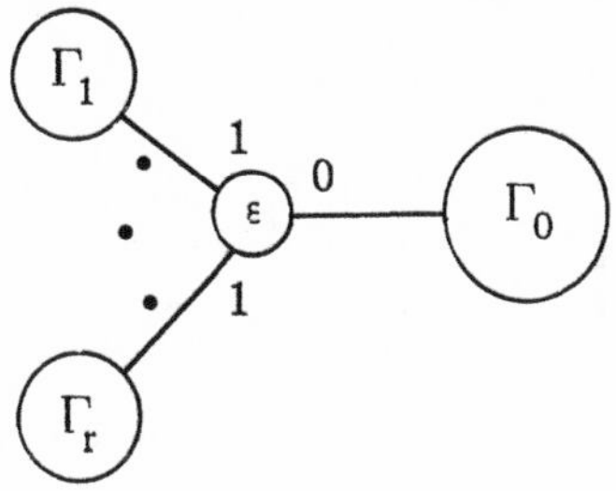

in the minimal splice diagram, the one arrowhead vertex belongs to Γ_0. Thus, for example, the simplest non-fiberable graph knot is perhaps

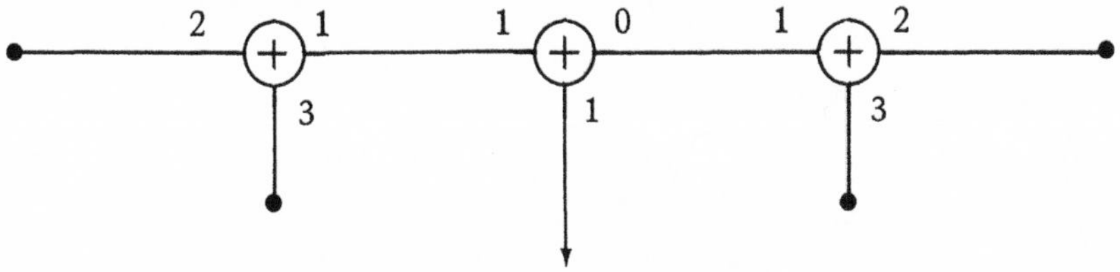

On the other hand, by Corollary 9.3, this cannot happen for solvable graph knots, that is, graph knots in S^3 :

COROLLARY 11.7. *Every solvable graph knot* (S^3,K) *is fiberable.*

12. *Alexander polynomial for graph links*

We return to the notations of section 10. That is, Γ is a splice diagram with arrowhead vertices $v_1,\cdots,v_n$ and remaining vertices $v_{n+1},\cdots,v_k$, representing a link $L = L(\Gamma) = (\Sigma, S_1 \cup \cdots \cup S_n)$. The following theorem computes the Alexander polynomial $\Delta^L = \Delta^L(t_1,\cdots,t_n)$. We assume $n \geq 1$, that is, the link is non-empty.

THEOREM 12.1. *Assume* Γ *is connected. Then, with* $\mathbf{L} = \mathbf{L}(\Gamma)$ *as above,*

$$\Delta^{\mathbf{L}}(t_1,\cdots,t_n) = \begin{cases} \prod_{j=n+1}^{k} (t_1^{\ell_{1j}} t_2^{\ell_{2i}} \cdots t_n^{\ell_{nj}} - 1)^{\delta_j - 2}, & n \geq 2 \\ (t_1 - 1) \prod_{j=2}^{k} (t_1^{\ell_{1j}} - 1)^{\delta_j - 2} & , n = 1 . \end{cases}$$

where the right-hand side should be interpreted as in the remark after 11.3, that is, terms of the form $(t_1^0 \cdots t_n^0 - 1)^a$ *should be formally cancelled against each other before being set equal to zero.*

In proving this theorem we shall also show:

THEOREM 12.2. *If* $\mathbf{L} = \mathbf{L}(\Gamma)$ *is as above then the following are equivalent:*

(i) $\Delta^{\mathbf{L}}(t_1,\cdots,t_n) = 0$

(ii) $\mathbf{L}$ *is algebraically split (that is: after reindexing if necessary, there exists a* q *with* $1 \leq q < n$ *and* $\ell(S_i, S_j) = 0$ *whenever* $1 \leq i \leq q < j \leq n$ *).*

(iii) $\mathbf{L}$ *is homologically split (that is: after reindexing as above, each multilink* $\mathbf{L}(m_1,\cdots,m_n)$ *on* $\mathbf{L}$ *has a Seifert surface which is a disjoint union of a Seifert surface for*

$$\mathbf{L}(m_1,\cdots,m_q,0,\cdots,0)$$

and one for

$$\mathbf{L}(0,\cdots,0,m_{q+1},\cdots,m_n) \text{).}$$

Moreover, these conditions imply the following condition (iv), and are equivalent to it if $\mathbf{L} = \mathbf{L}(\Gamma)$ *is a link in* S^3 *(solvable link).*

(iv) *No multilink* $\mathbf{L}(\underline{m})$ *on* $\mathbf{L}$ *is fibered.*

One might guess that the analogous results to 12.1 and 11.3 hold always for the single variable Alexander polynomial $\Delta^{\mathbf{L}(\underline{m})}$ of a multilink on $\mathbf{L}$; however the interpretation of how zero terms cancel in this case needs more care (the case that there are no zero terms is just Theorem 11.3). For example, the toral multilink given by

$\Gamma(\underline{m}) =$

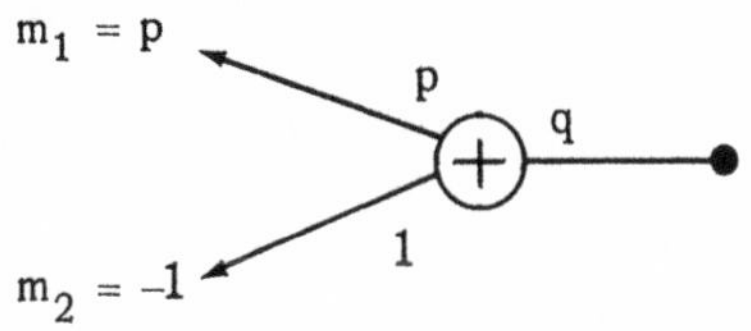

has 1-variable Alexander polynomial obtained by specializing

$$\Delta^{\mathbf{L}} = (t_1^q t_2^{pq} - 1)/(t_1 t_2^p - 1)$$

at $t_1 = t^p, t_2 = t^{-1}$ and then multiplying by $t-1$ (Theorem 5.1). This looks like $(t-1)(t^0-1)/(t^0-1)$, but since $\Delta^{\mathbf{L}} = (t_1 t_2^p)^{q-1} + \cdots + (t_1 t_2^p) + 1$, the answer is $\Delta^{\mathbf{L}(\underline{m})} = qt-q$, not $\Delta^{\mathbf{L}(\underline{m})} = t-1$.

The proof of 12.1 is in principle a direct consequence of the Seifert case plus the description in Theorem 5.3 of how $\Delta^{\mathbf{L}}$ behaves under splicing. In practice the cancelling zero factors necessitate care in carrying out the proof, which is an induction on the size of Γ.

Proofs. Suppose first that no vertex v_j of Γ with $n+1 \leq j \leq k$ satisfies $(\ell_{1j}, \cdots, \ell_{nj}) = (0, \cdots, 0)$. Then, for most multiplicities $\underline{m} = (m_1, \cdots, m_n)$, $\mathbf{L}(\underline{m})$ is fibered and the characteristic polynomial $\Delta_1^{\mathbf{L}(\underline{m})}$ of the monodromy is the specialization $t_i = t^{m_i}$ of $\Delta^{\mathbf{L}}(t_1, \cdots, t_n)$ multiplied by $t^d - 1$, by 11.2 and 5.1. Since we know $\Delta_1^{\mathbf{L}(\underline{m})}$ by 11.3, this forces the claimed value for $\Delta^{\mathbf{L}}(t_1, \cdots, t_n)$, proving Theorem 12.1 for this case.

We must thus just consider the case that vertices with $n+1 < j \leq k$ and $(\ell_{1j}, \cdots, \ell_{nj}) = (0, \cdots, 0)$ do occur. We call such a vertex a *zero-vertex*. We first show that such a zero vertex can be chosen in a very special way.

Claim: Γ contains a zero-vertex v_j as follows, such that Γ_0 contains no arrowheads:

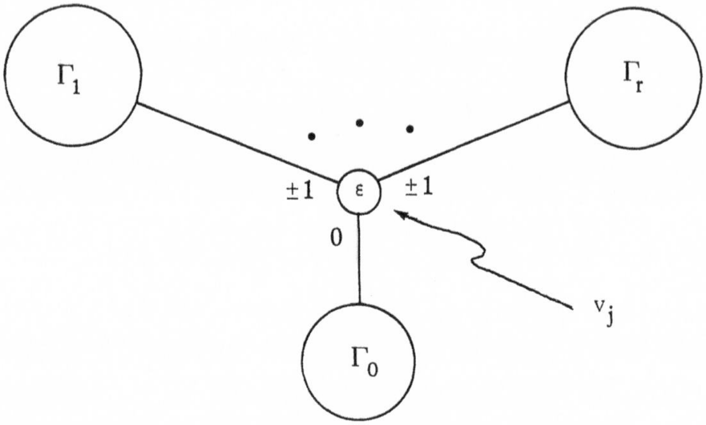

Indeed, if we do not require that Γ_0 contain no arrowheads, such a v_j is easily found: the closest zero-vertex to any arrowhead of Γ must look like this, since otherwise the next closer vertex to the arrowhead would still be a zero-vertex. Now if Γ_0 has an arrowhead, the path from v_j to that arrowhead will pass through a vertex of the same type as v_j but with a smaller Γ_0. Thus, if we choose our v_j with Γ_0 as small as possible, Γ_0 will contain no arrowheads, and the claim is proved.

We now assume v_j is chosen as in the above claim. After reindexing, if necessary, we may assume $\Gamma_1, \cdots, \Gamma_s$ contain arrowheads and $\Gamma_{s+1}, \cdots, \Gamma_r$ do not, with $1 \leq s \leq r$.

Case 1: $s = 1$. In this case every vertex of $\Gamma_2 \cup \cdots \cup \Gamma_r$ is a zero-vertex. Let Γ' be the diagram

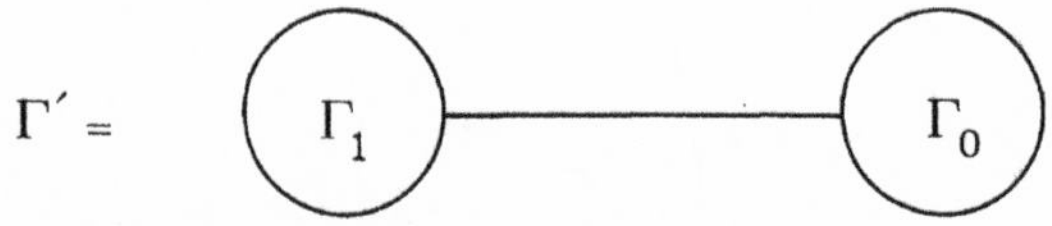

obtained by reducing, by Theorem 8.1, 3) the diagram

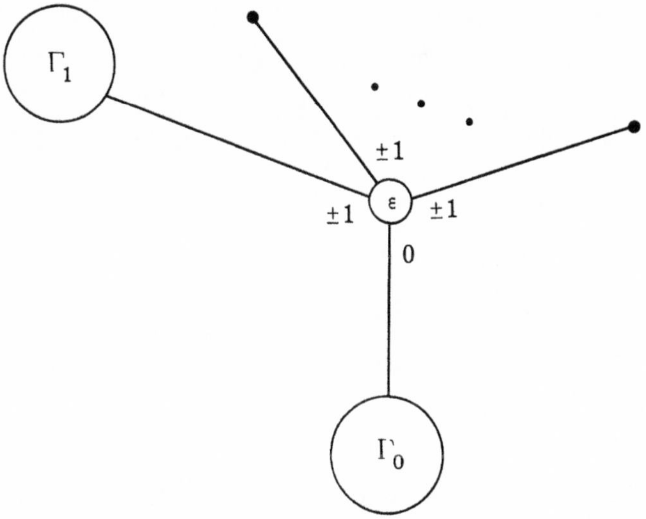

By Theorem 5.3 $\mathbf{L}(\Gamma)$ and $\mathbf{L}(\Gamma')$ have identical Alexander polynomials. On the other hand the formula of 12.1 applied to Γ differs from the same formula applied to Γ' only by terms of the form $(t_1^0 \cdots t_n^0 - 1)^{\delta_i - 2}$. The total exponent of these terms is $\Sigma(\delta_i - 2)$, summed over

$$\{i \mid v_i \in \{v_j\} \cup \operatorname{vert}(\Gamma_2 \cup \cdots \cup \Gamma_r)\},$$

and a simple induction shows this sum to yield zero. Thus, the validity of 12.1 for Γ follows from its validity for the smaller diagram Γ'.

Case 2: $s \geq 2$. In this case we call v_j a *splitting vertex* of Γ. A simple counting argument shows that the total exponent of terms of the form $(t_1^0 \cdots t_n^0 - 1)^{\delta_i - 2}$, in the formula of 12.1, is at least $s - 1$. Thus, the proof of 12.1 will be completed by showing that $\Delta^{\mathbf{L}} = 0$ in this case. In fact, we shall simultaneously prove most of 12.2:

LEMMA 12.3. *If Γ is connected, the existence of a splitting vertex in Γ is equivalent to each of conditions i) to iii) of 12.2.*

Proof. We have just seen that the only way $\Delta^{\mathbf{L}}$ can be zero is that Γ contain a splitting vertex (the existence or non-existence of a splitting vertex in Γ is preserved in the inductive step of "Case 1" above). Thus, we have proved implication (1) of the following diagram of implications.

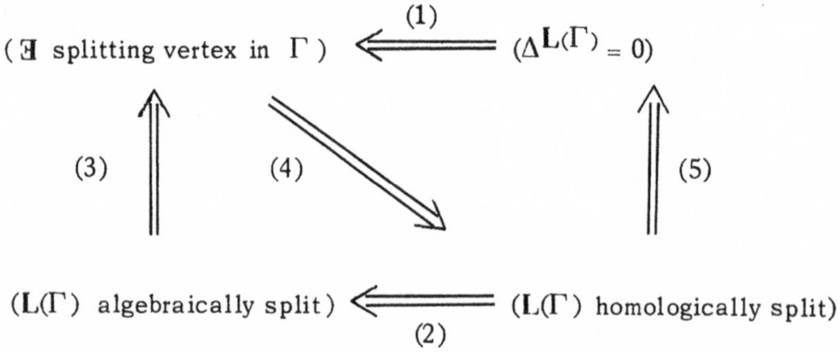

Implication (2) is trivial. For (3) let $\{v_1,\cdots,v_n\} = X_1 \cup \cdots \cup X_m$ be the finest partition of $\{v_1,\cdots,v_n\}$ for which $\ell(S_i, S_j) = 0$ whenever v_i and v_j are in distinct members of the partition. If $\Gamma^{(i)}$ is the smallest connected subgraph of Γ containing X_i, then $\Gamma^{(i)}$ will be disjoint from $\Gamma^{(j)}$ for $i \neq j$. The path in Γ connecting $\Gamma^{(1)}$ to $\Gamma^{(2)}$ must pass through a node with an adjacent zero edge weight:

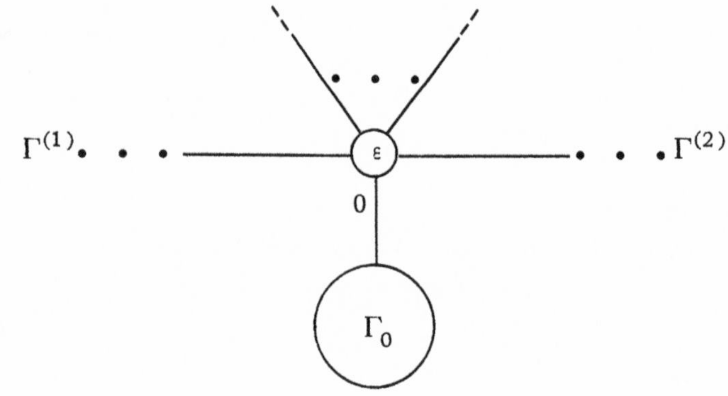

If this is not a splitting vertex, that is, if Γ_0 contains arrowheads, our earlier argument to find the "special" zero vertex v_j discovers a splitting vertex in Γ_0.

To see implication (4) assume Γ is as described earlier, with splitting vertex v_j. Then $L(\Gamma)$ is the result of splicing links

$$L_i = L(\Gamma_i'), \quad i = 0, \cdots, r+1 ,$$

as follows

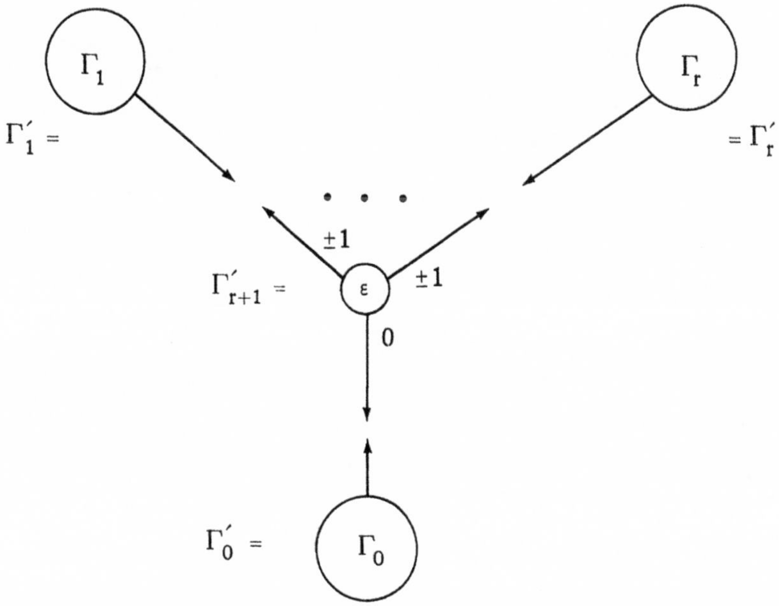

Let L' be the result of splicing $L_1, \cdots, L_{r+1}$, so L' is a connected sum of $L_1, \cdots, L_r$, along components $S_1, \cdots, S_r$ say. Let S_0 be the summed component in L'.

If $L(\underline{m})$ is a multilink on $L = L(\Gamma)$, let $L_i(\underline{m}_i)$ and $L'(\underline{m}')$ be the corresponding multilink splice summands. By 10.6, S_i has multiplicity 0 in $L_i(\underline{m}_i)$ for $i = 1, \cdots, r$, so a Seifert surface for $L_i(\underline{m}_i)$ will be a surface which appears to intersect S_i transversally in several points. Thus in forming L' we can take connected sum of the L_i away from these points on the S_i, obtaining a disconnected Seifert surface for $L'(\underline{m}')$. Actually it may not be an "honest" Seifert surface since it may meet the boundary of a tubular neighborhood of S_0 in curves which could be homologically cancelled. But in any case, in splicing L' to L_0 to form L we can now paste this surface in L' to a collection of parallel Seifert surfaces of L_0 to obtain the desired disconnected Seifert surface for $L(\underline{m})$.

Finally, implication (5) follows by Proposition 5.1 and the following general fact.

PROPOSITION 12.4. *Let* $L(\underline{m}) = (\Sigma, m_1 S_1 \cup \cdots \cup m_n S_n)$ *be any multilink which admits a Seifert surface* F *(with no closed components) whose closure is disconnected. Then the single variable Alexander polynomial satisfies* $\Delta^{L(\underline{m})} = 0$.

Proof. Suppose F is the disjoint union of F_1 and F_2. Choose a short transverse arc to each of F_1 and F_2. By connectedness of $\Sigma - F$ we can connect these two arcs to get a closed curve K in the link exterior Σ_0 such that K intersects each of F_1 and F_2 transversally in one point, but with opposite orientations, so its total intersection number with F is zero. Thus $\underline{m}(K) = 0$, so K lifts to an infinite collection $\cdots, K_{-1}, K_0, K_1, \cdots$ of closed curves in the infinite cyclic cover $\Sigma_0(\underline{m})$ of Σ_0 determined by $\underline{m}$. Now F_1 lifts to an infinite number of copies $F_1^{(i)}$, $i \in Z$ in $\Sigma_0(\underline{m})$, and we can index them such that the intersection number $K_i \cdot F_1^{(j)}$ is the Kronecker delta δ_{ij}. Thus, the K_i are linearly independent in $H_1(\Sigma_0(\underline{m}))$ and generate a submodule isomorphic to $Z[t, t^{-1}]$. But $\Delta^{L(\underline{m})}$ is the greatest common divisor of the order ideal of $H_1(\Sigma_0(\underline{m}))$ (Proposition 5.1. ii), and is thus zero.

To complete the proof of Theorem 12.2 we must just show that if $L(\Gamma)$ is a link in S^3 and if no multilink $L(\Gamma(\underline{m}))$ is fibered, then Γ has a splitting vertex. Suppose the contrary. Then we may suppose Γ is a minimal splice diagram and is as in "Case 1" of the proof of 12.1

above. As in section 9, after replacing all arrowheads of Γ by ordinary vertices, Γ reduces to a trivial graph by the moves of Theorem 8.1. But these reduction moves can never affect $\Gamma_0, \Gamma_2, \cdots, \Gamma_r$, since these portions of Γ already contain no arrowheads and Γ is minimal. This is clearly a contradiction, so it completes the proof.

13. *Mondromy revisited*

Let $L(\Gamma(\underline{m}))$ be a fibered graph multilink. Let F be the fiber and $h: F \to F$ the monodromy. By Theorem 4.2 and its addendum we may assume $h: F \to F$ is pieced together from the monodromy maps $h_v: F_v \to F_v$ of the fibers of the splice components of $L(\Gamma(\underline{m}))$. We assume $\Gamma(\underline{m})$ is a minimal splice diagram, so the index v here can be taken to run over the nodes of $\Gamma(\underline{m})$. By Lemma 11.4, h_v is isotopic to a periodic homeomorphism of order $|\underline{m}(S_v)|$, where S_v is the "virtual component" corresponding to node v (that is a nonsingular fiber of the corresponding Seifert component; in our earlier notation we wrote S_i for S_v with $v = v_i$).

By a suitable isotopy of h and a slight change of notation we may assume that the pieces F_v of F are separated by annuli in F and that $h_v = h|F_v$ itself has finite order on F_v. An example is given in the following picture.

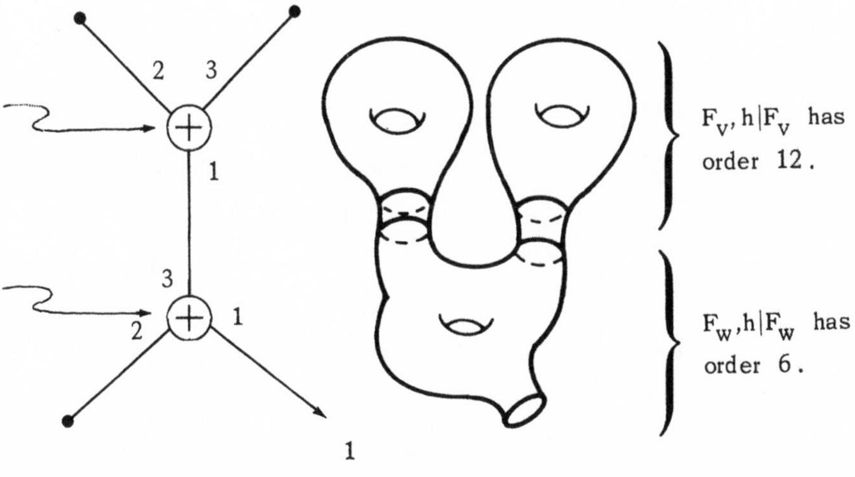

To complete the description of h, we must describe the action of h on the separating annuli.

Let q be a common multiple of the orders of the h_i, so h^q is the identity on each F_v. Then h^q acts as a "twist map" on each annulus. The topological description of h will be completed by computing the amount of twist.

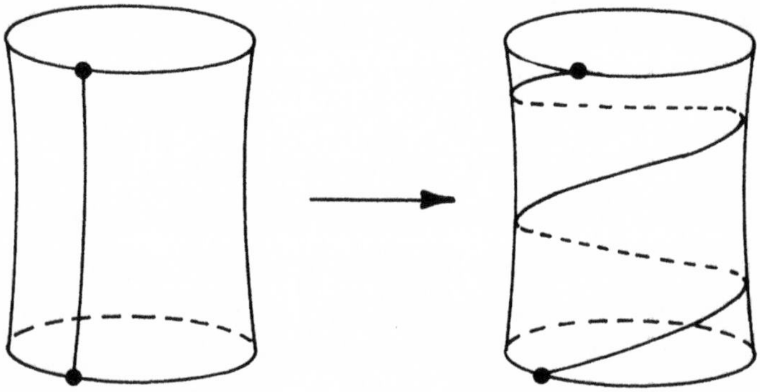

DEFINITION. Let $\phi : A \to A$ be a homeomorphism of the oriented annulus $A \cong S^1 \times [0,1]$ with $\phi|\partial A = \mathrm{id}$. The twist of ϕ is defined as

$$\mathrm{twist}(\phi) = X \cdot \mathrm{var}_\phi(X) \qquad \text{(intersection number)},$$

where $X \in H_1(A, \partial A; \mathbf{Z})$ is a generator and

$$\mathrm{var}_\phi : H_1(A, \partial A : \mathbf{Z}) \to H_1(A; \mathbf{Z})$$

is defined by $\mathrm{var}_\phi[C] = [\phi(C) - C]$, for any cycle C.

More generally if B is a disjoint union of annuli and $h : B \to B$ a homeomorphism with $h^q|\partial B = \mathrm{id}$ for some $q > 0$, define, for any component A of B,

$$\mathrm{twist}(h|A) = \frac{1}{q}\,\mathrm{twist}(h^q|A).$$

In the case at hand, B is the collection of separating annuli in the fiber F of our multilink $L(\Gamma(\underline{m}))$. Each edge connecting two nodes of $\Gamma(\underline{m})$ will correspond to a collection of components of B which are cyclically permuted by h and which, therefore, yield the same twist. The twist thus associates a rational number t_E to each edge E of Γ which connects two nodes.

Let v and w be two adjacent nodes of $\Gamma(\underline{m})$ as in the illustration:

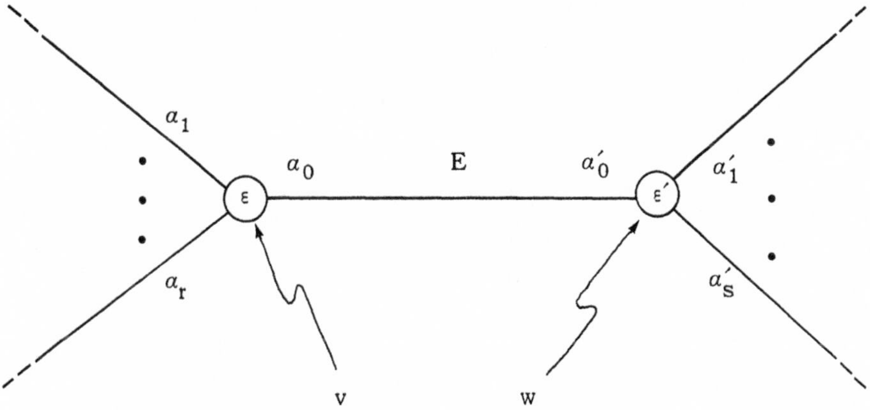

Let E be the edge connecting them. Let m and m′ be as in Corollary 10.6 and let $d_E = \gcd(m, m')$. Let $\ell = \underline{m}(S_v), \ell' = \underline{m}(S_w)$.

THEOREM 13.1. *There are exactly* d_E *annuli in* F *corresponding to edge* E. *If* A *is any one of them, then*

$$t_E = \text{twist}(h|A) = \frac{-d_E}{\ell \cdot \ell'}(\alpha_0\alpha_0' - \varepsilon\varepsilon'\alpha_1 \cdots \alpha_r\alpha_1' \cdots \alpha_s') .$$

Proof. There is another way of thinking of the twist which will be useful in computing it. By taking tubular neighborhoods of the separating tori in the link exterior we can express the link exterior as a disjoint union of Seifert pieces joined together by toral annuli $T^2 \times I$. By Lemma 11.4 and its proof, the monodromy of a Seifert link is realized by the flow along the fibers of the Seifert fibration of the link exterior. The monodromy of an arbitrary fibered graph multilink is thus realized by a "monodromy flow", which on each Seifert piece just flows along the Seifert fibers, while on a connecting toral annulus $T^2 \times I$ it interpolates between the flows on $T^2 \times \{0\}$ and $T^2 \times \{1\}$ given by the Seifert fibrations at each end of $T^2 \times I$. The twist measures the difference between these two flows. This is the approach we shall take to compute it.

We use the notation of Theorem 10.5 and Corollary 10.6, so

$$L(\Gamma(\underline{m})) = (\Sigma, m_1 S_1 \cup \cdots \cup m_n S_n)$$

is the result of splicing two multilinks with splice diagrams

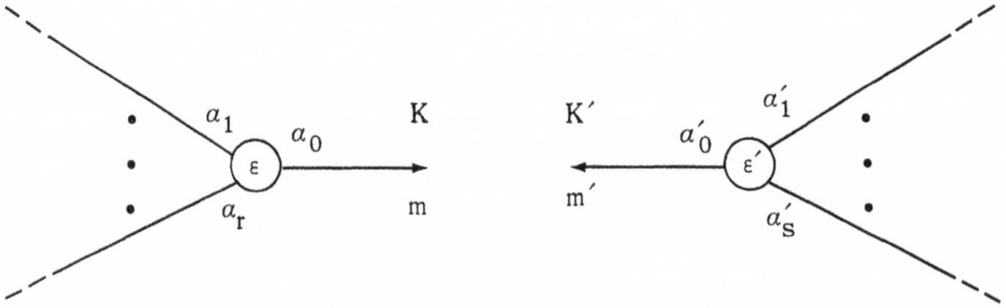

This splicing is along a torus T in Σ. In T we have curves $M = L'$, $L = M'$, which are respectively meridian and longitude of K (longitude and meridian of K'). We also have fibers H and H' in T of the Seifert structures to the left and right of T.

We write the fiber of $L(\Gamma(\underline{m}))$ as $F = F_0 \cup F_0'$, where F_0 and F_0' are the fibers of the two splice summands, so $F_0 \cap F_0' = F \cap T$. We orient $F \cap T$ as part of ∂F_0, or equivalently as part of $-\partial F_0'$. Then in homology of T we have a relation

$$F \cap T = mL - m'M .$$

Thus $F \cap T$ consists of $d_E = \gcd(m, m')$ components, each homologous to $(m/d_E)L - (m'/d_E)M$.

We shall need some further formulae. Denote

$$a = \varepsilon a_1 \cdots a_r , \quad a' = \varepsilon' a_1' \cdots a_s' .$$

Then, as in the proof of 8.4, we have

$$(13.2)\qquad \begin{cases} H = \alpha_0 L + \alpha M \\ H' = \alpha' L + \alpha_0' M\,. \end{cases}$$

Actually, in 8.4 this was only proved when $\varepsilon = \varepsilon' = 1$. The general case of 13.2 (or indeed of 13.1) follows from the case $\varepsilon = \varepsilon' = 1$ using part 2 of Theorem 8.1. Theorems 10.1 and 10.6 give:

$$(13.3)\qquad \begin{cases} \ell = \alpha m + \alpha_0 m', \\ \ell' = \alpha_0' m + \alpha' m'. \end{cases}$$

We shall adjust the velocity of the monodromy flow so that the monodromy $h: F \to F$ is the time 1 map of the flow. We have on T two flows, coming from the Seifert structures to each side of T, and the desired twist is determined by the difference between these flows.

We coordinatize T as $T = \mathbf{R}/\mathbf{Z} \times \mathbf{R}/\mathbf{Z}$, with $L = \mathbf{R}/\mathbf{Z} \times \{0\}$ and $M = \{0\} \times \mathbf{R}/\mathbf{Z}$. By 11.4 and its proof, the monodromy flow coming from the Seifert structure to the left of T runs parallel to fiber H at speed $1/\ell$ (taking unit speed to be one circuit per unit time). By (13.2) this is the flow

$$(\alpha_0/\ell)\left(\frac{\partial}{\partial x_1}\right) + (\alpha/\ell)\left(\frac{\partial}{\partial x_2}\right).$$

Similarly the other flow on T runs parallel to H' at speed $1/\ell'$, so it is the flow

$$(\alpha'/\ell')\left(\frac{\partial}{\partial x_1}\right) + (\alpha_0'/\ell')\left(\frac{\partial}{\partial x_2}\right).$$

The difference between these flows is

$$\left(\frac{\alpha'}{\ell'} - \frac{\alpha_0}{\ell}\right)\left(\frac{\partial}{\partial x_1}\right) + \left(\frac{\alpha_0'}{\ell'} - \frac{\alpha}{\ell}\right)\left(\frac{\partial}{\partial x_2}\right),$$

which simplifies, using (13.3), to

$$\frac{-d_E}{\ell \cdot \ell'}(\alpha_0 \alpha_0' - \alpha\alpha')\left[\frac{m}{d_E}\frac{\partial}{\partial x_1} - \frac{m'}{d_E}\frac{\partial}{\partial x_2}\right].$$

Since the flow in square brackets is the flow which circuits a component of $F \cap T$ in unit time, the theorem is proved.

Our picture of the monodromy $h: F \to F$ is still not quite complete, since for many purposes one wishes to assume that the monodromy is induced from a monodromy flow on the ambient homology sphere which follows meridians near any link component. This specifies $h|\partial F$, so one must interpolate twist maps on annular collar neighborhoods of the boundary components.

Suppose the multilink component in question is $m_1 S_1$:

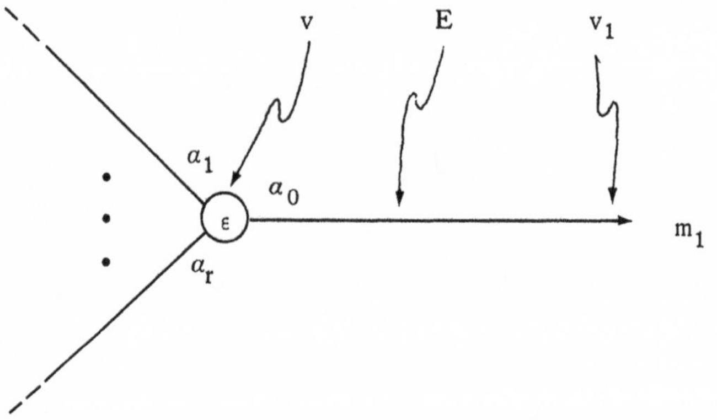

With notation as in the diagram, denote

$$\ell = \underline{m}(S_v), \quad m' = \ell(S_1, m_2 S_2 + \cdots + m_n S_n), \quad d_E = \gcd(m_1, m').$$

THEOREM 13.5. *F has* d_E *boundary components near* S_1. *If* A *is an annular neighborhood of one of these boundary components then*

$$\text{twist}(h|A) = \frac{-d_E}{m_1 \ell}\alpha_0$$

The proof is as for (13.1), hence omitted.

Example. The twists in the earlier example, using Theorems 13.1 and 13.5, are as follows:

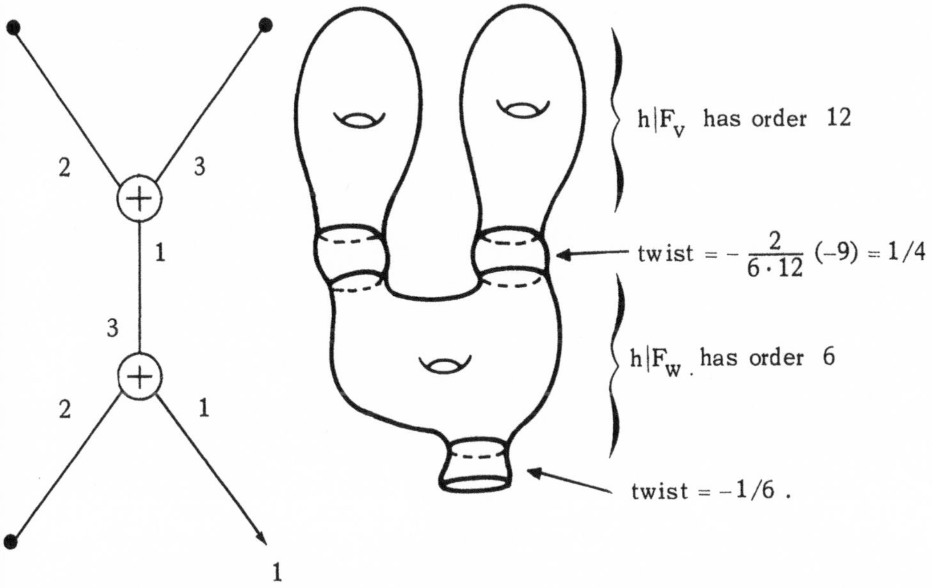

As a consequence of our geometric description of $h: F \to F$ we have the following theorem, which generalizes a result of Grothendieck for complex plane curve singularities.

THEOREM 13.6. *Let* q *be a multiple of the orders* $|\underline{m}(S_i)|$ *of the finite order pieces* $h_v: F_v \to F_v$ *of* $h: F \to F$. *Then the algebraic monodromy* $h_* = H_1(h): H_1(F) \to H_1(F)$ *satisfies* $(h_*^q - 1)^2 = 0$.

Proof. $h_*^q - 1$ applied to any homology class x will be a linear combination of the homology classes of the circles in F which separate the F_i (this is clear geometrically, but will be computed more precisely in the proof of Theorem 14.1). Since h_*^q fixes any such circle, $h_*^q(h_*^q - 1)x = (h_*^q - 1)x$, so $(h_*^q - 1)^2 x = 0$.

14. *Jordan normal form of the algebraic monodromy*

Let $L(\Gamma(\underline{m}))$ be a fibered graph multilink with fiber F and monodromy $h: F \to F$, as in the previous section. Theorem 11.3 may be regarded as telling us the eigenvalues of the algebraic monodromy $h_1: H_1(F) \to H_1(F)$. Theorem 13.6 tells us that the Jordan normal form of h_* consists only of 1×1 and 2×2 Jordan blocks. In this section we shall compute the characteristic polynomial of $h_*|(h_*^q - 1)(H_1(F))$, where q is as in Theorem 13.6. Since the roots of this polynomial, with multiplicities, will be the eigenvalues of h_* that appear in 2×2 Jordan blocks of h_*, this is equivalent to giving the Jordan normal form of h_*.

Let $\mathfrak{E}$ be the set of edges E of Γ which connect two nodes. For each $E \in \mathfrak{E}$ we denoted by d_E the number of components of $F \cap T_E$ where $T_E \subset \Sigma$ is the separating torus corresponding to edge E. Recall from Theorem 13.1 that $d_E = \gcd(m, m')$ with m and m′ as in Corollary 10.6.

Let $\mathfrak{N}$ be the set of nodes of Γ. For $v \in \mathfrak{N}$ let F_v be the fiber of the corresponding Seifert multilink splice component. Let d_v be the number of components of F_v. By the remarks preceding Theorem 4.3, d_v is the g.c.d. of the link component multiplicities of the corresponding Seifert splice component. In practice it is simpler to compute d_v as the g.c.d. of all d_E's of edges $E \in \mathfrak{E}$ which meet v together with all m_i's of arrowheads v_i of Γ adjacent to v.

The number d of components of F, namely $d = \gcd(m_1, \cdots, m_n)$, is also equal to the g.c.d. of the d_v, $v \in \mathfrak{N}$.

DEFINITION. We say $L(\Gamma(\underline{m}))$ *has uniform twists* if the twists t_E, $E \in \mathfrak{E}$, as computed in Theorem 13.1, are either all positive or all negative.

In section 23 we show that the multilinks which arise in algebraic geometry are of this type; in fact, they have only negative twists (see Theorem 9.4 for the case of links in S^3).

THEOREM 14.1. *Let* q *be as in Theorem 13.6.*

i) *If* $L(\Gamma(\underline{m}))$ *has uniform twists, then the characteristic polynomial* $\Delta'(t)$ *of* $h_*|(h_*^q - 1)(H_1(F))$ *is equal to*

$$p(t) = (t^d - 1) \prod_{E \in \mathcal{E}} (t^{d_E} - 1) / \prod_{v \in \mathcal{N}} (t^{d_v} - 1) .$$

ii) *If* $L(\Gamma(\underline{m}))$ *does not have uniform twists,* $\Delta'(t)$ *is a quotient of* p(t) *by a (usually trivial) polynomial described at the end of the proof.*

iii) *Let* $\mathcal{N}'$ *be the set of separating nodes of* Γ, *that is, nodes* v *of* Γ *for which at least two components of* $\Gamma - \{v\}$ *have arrowheads. Let* $\mathcal{E}'$ *be the set of separating edges, that is, edges connecting separating nodes. Then*

$$p(t) = \begin{cases} (t^d - 1) \prod_{E \in \mathcal{E}'} (t^{d_E} - 1) / \prod_{v \in \mathcal{N}'} (t^{d_v} - 1), & (\mathcal{N}' \neq \emptyset) , \\ 1 & , (\mathcal{N}' = \emptyset) . \end{cases}$$

Part iii) shows that every fiberable graph multiknot (and thus, by Corollary 11.7, every solvable knot in S^3) has finite order algebraic monodromy, and gives the criterion

$$d + \Sigma_{E \in \mathcal{E}'} d_E = \Sigma_{v \in \mathcal{N}'} d_v$$

for the finiteness of the monodromy of a graph multilink with more than one component. In the case of a link, rather than a multilink, we can do better.

COROLLARY 14.2. *Suppose* $L(\Gamma(\underline{m}))$ *has uniform twists. Assume* $m_i = \pm 1$ *for all* i *(that is,* $L(\Gamma(\underline{m})$ *is a link rather than a multilink). Then the algebraic monodromy has finite order if and only if* $d_E = 1$ *for every separating edge* E.

Proof (14.1) As in section 13, we write F as the union of pieces F_v, $v \in \mathfrak{N}$, on which h has finite order, and collections B_E, $E \in \mathfrak{E}$ of annuli connecting these pieces. Denote

$$B = \bigcup_{E \in \mathfrak{E}} B_E, \quad G = \bigcup_{v \in \mathfrak{N}} F_v,$$

and consider the following diagram with exact rows:

$$(14.2) \qquad \begin{array}{ccccc}
H_1(F) & \xrightarrow{\ i\ } & H_1(F,G) & \longrightarrow & H_0(G) \\
\Big\downarrow{\scriptstyle h_*^q - 1} & & \Big\downarrow{\scriptstyle T} & \searrow{\scriptstyle e}\,\cong & \\
 & & & & H_1(B,\partial B) \\
 & & & \swarrow{\scriptstyle \mathrm{var}} & \\
H_1(F) & \xleftarrow{\ j\ } & H_1(B) & \xleftarrow{\ \partial\ } & H_2(F,B)
\end{array}$$

Here e is an excision isomorphism, var is $\mathrm{var}\, h^q$, and T is defined as the composite $(\mathrm{var}) \cdot e$. The commutativity of the diagram is clear from the definition of var. Thus, $\mathrm{Im}(h_*^q - 1) = j\,T(\mathrm{Im}(i))$.

CLAIM. If $L(\Gamma(\underline{m}))$ has uniform twists, then $j\,T|\mathrm{Im}(i)$ is injective, so $\mathrm{Im}(h_*^q - 1) \cong \mathrm{Im}(i)$.

Indeed, assume uniform twists. For $x, y \in H_1(F,G)$ define $\langle x, y\rangle = x \cdot Ty$ (intersection number). The matrix of this form with respect to the natural basis of $H_1(F,G) = H_1(B,\partial B)$ is a diagonal matrix D with d_E entries qt_E for each edge $E \in \mathfrak{E}$. It is thus either positive or negative definite. Now for $x \in H_1(F)$ with $i(x) \neq 0$ we have $x \cdot (jT\,i(x)) = i(x) \cdot T\,i(x) = \langle i(x), i(x)\rangle$, which is nonzero by definiteness of this form. Thus, $i(x) \neq 0$ implies $jT\,i(x) \neq 0$, which was our claim.

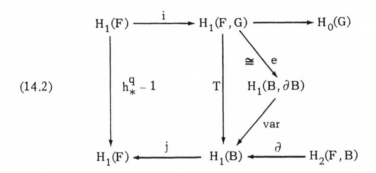

Now h induces automorphisms of all groups in diagram (14.2) and all maps are compatible with these automorphisms, so our isomorphism $\mathrm{Im}(h_*^q - 1) \cong \mathrm{Im}(i)$ is compatible with the induced automorphisms. The exact sequence for the pair (F, G) gives an exact sequence

$$\begin{array}{c} 0 \to \mathrm{Im}(i) \to H_1(F, G) \to H_0(G) \to H_0(F) \to 0 . \\ \| \\ H_1(B, \partial B) \end{array}$$

The characteristic polynomials for h_* on the three groups $H_1(B, \partial B)$, $H_0(G)$, and $H_0(F)$ are evidently

$$\prod_{E \in \mathcal{E}} (t^{d_E} - 1) , \quad \prod_{v \in \mathcal{N}} (t^{d_v} - 1) , \quad (t^d - 1)$$

respectively, so the characteristic polynomial of

$$h_* | \mathrm{Im}(i) \cong h_* | \mathrm{Im}(h_*^q - 1)$$

is as claimed in the theorem.

Even if $L(\Gamma(\underline{m}))$ does not have uniform twists, the map T is given by the diagonal matrix D and is hence bijective. The diagram (14.2) yields the isomorphism

$$\mathrm{Im}(h_*^q - 1) \cong \mathrm{Im}(i)/(T^{-1}(\mathrm{Im}(\partial)) \cap \mathrm{Im}(i)) ,$$

so $\Delta'(t)$ is the same polynomial as before, divided now by the characteristic polynomial of $h_* | (T^{-1}(\mathrm{Im}\, \partial) \cap \mathrm{Im}(i))$. This is easy to compute explicitly in any concrete case, because of the explicitness of the maps involved. One can also give a "closed formula" for it – this was done in our original version [E-N] – but the gain is not really worth the effort.

To prove (iii), suppose we have $E \in \mathcal{E} - \mathcal{E}'$ as follows,

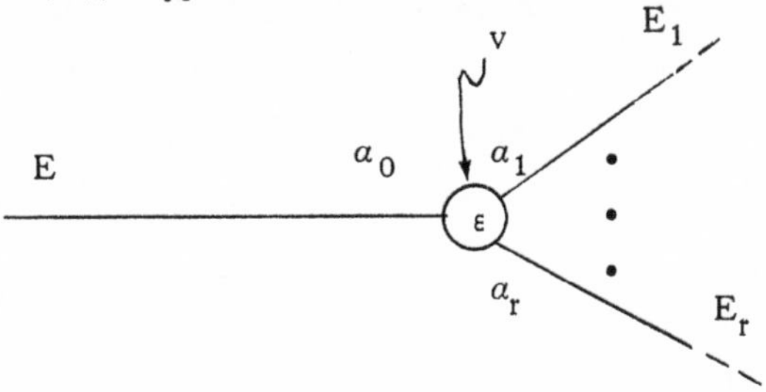

,

such that all arrowheads of Γ are to the left of E. Then each d_{E_i} is a multiple of d_E, namely $d_{E_i} = \alpha_1 \cdots \hat{\alpha}_i \cdots \alpha_r d_E$, by Corollary 10.6. Hence, $d_v = d_E$. Thus, starting at outermost edges in $\mathcal{E}$ we can remove edges of $\mathcal{E} - \mathcal{E}'$ and vertices in $\mathcal{N} - \mathcal{N}'$ one by one in the formula for p(t), without changing p(t), so (iii) is proved.

Proof (14.2). The "if" is clear, and is true even without the uniform twists condition. To see the "only if", let v be an extremal vertex of $\mathcal{N}'$ (that is, one with only one edge E of $\mathcal{E}'$ abutting). Then deleting the contributions of v and E from the formula for p(t) decreases the degree of p(t) by the nonnegative quantity $d_E - d_v$. Repeating this iteratively, we can delete vertices until just one $v \in \mathcal{N}'$ remains, with $d_v = 1$. This process reduces p(t) to $p(t) = 1$. But, by assumption, p(t) started as $p(t) = 1$. Thus, p(t) remained unchanged at each step of the process, so $d_E = d_v$ at each stage, so all d_E and d_v must have equalled 1.

15. *Spliced knots: Seifert forms and algebraic monodromy*

In the case of *knots* (Σ, S) we can give closed computations of some finer invariants, in particular of the Seifert form. These results are not restricted to graph knots, and apply to some links of several components also.

If $L(\underline{m}) = (\Sigma, m_1 S_1 \cup \cdots \cup m_n S_n)$ is a multilink and F a Seifert surface, then the *Seifert form* is the form

$$A : H_1(F; Z) \otimes H_1(F; Z) \to Z$$

defined by $A(x,y) = \ell(x_+, y)$, where x_+ is the result of translating x into $\Sigma - F$ in the positive normal direction. A depends on the choice of Seifert surface: what Seifert surface is being used will be clear from context.

A Seifert surface for $L(k\underline{m})$ can be taken to be $|k|$ parallel copies of F, with orientation given by the sign of k. It follows then directly from the definition that the corresponding Seifert form is:

$$A^{(k)} = \begin{cases} \left.\begin{pmatrix} A & A^t & A^t \cdots A^t \\ A & A & A^t \cdots A^t \\ \cdot & & \cdot \\ & & \cdot \\ \cdot & & \cdot \\ A & A & A \cdots A \end{pmatrix}\right\} k, & \text{if } k > 0 \\ \left.\begin{pmatrix} A^t & A & A \cdots A \\ A^t & A^t & A \cdots A \\ \cdot & & \cdot \\ \cdot & & \\ \cdot & & \cdot \\ A^t & A^t & A^t \cdots A^t \end{pmatrix}\right\} |k|, & \text{if } k < 0 . \end{cases}$$

Moreover if $L(\underline{m})$ is fibered, and H is a matrix for its monodromy $H_1(F) \to H_1(F)$ then $L(k\underline{m})$ clearly has algebraic monodromy:

$$H^{(k)} = \left.\begin{pmatrix} 0 & 0 \cdots 0 & H^{\varepsilon} \\ I & 0 \cdots 0 & 0 \\ 0 & I \cdots 0 & 0 \\ & & \\ 0 & 0 \cdots I & 0 \end{pmatrix}\right\} |k|, \quad \varepsilon = \operatorname{sign}(k) .$$

Now suppose $L(\underline{m}) = (\Sigma, m_1 S_1 \cup \cdots \cup m_n S_n)$ is the result of splicing $L'(\underline{m}') = (\Sigma', m_0' S_0')$ to $L''(\underline{m}'') = (\Sigma'', m_0'' S_0'' \cup m_1 S_1 \cup \cdots \cup m_n S_n)$ along S_0', S_0''. Let $L_0''(\underline{m}) = (\Sigma'', m_1 S_1 \cup \cdots \cup m_n S_n)$. For graph links a typical example might be

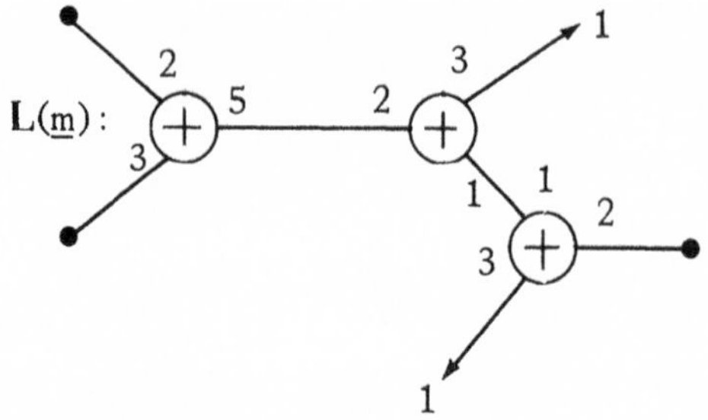

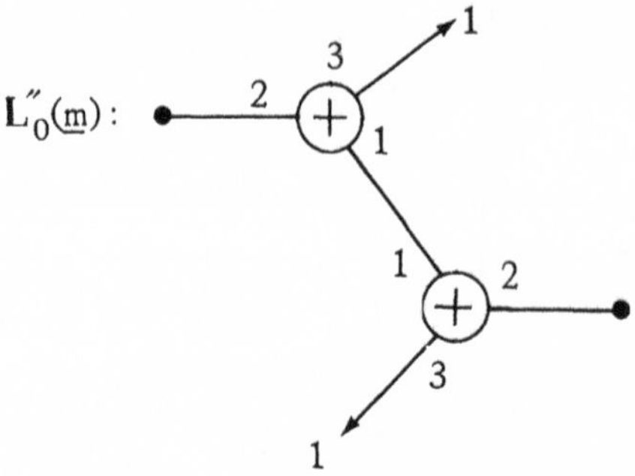

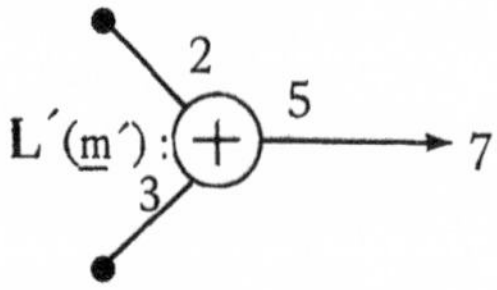

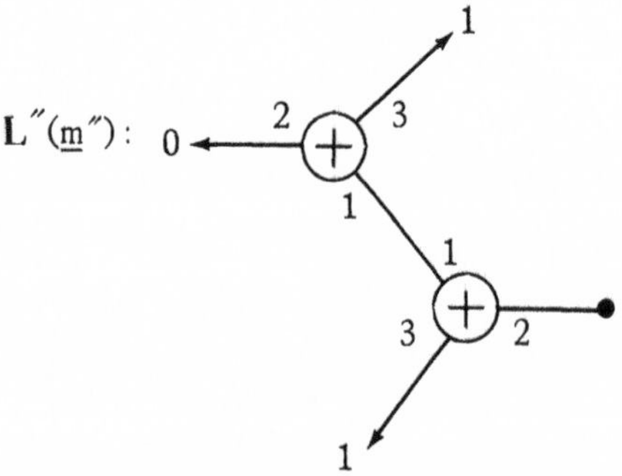

A Seifert surface F for $L(\underline{m})$ can be obtained by gluing together Seifert surfaces F' and F'' for $L'(\underline{m}')$ and $L''(\underline{m}'')$. By capping boundary components of F'' with discs we get a Seifert surface F''_0 for $L''_0(\underline{m})$. The Mayer Vietoris sequence

$$H_1(F' \cap F'') \to H_1(F') \oplus H_1(F'') \to H_1(F) \to 0$$

implies

$$H_1(F) = H_1(F') \oplus H_1(F''_0) ,$$

since $F' \cap F'' = \partial F'$. This isomorphism is compatible with Seifert linking forms, so we get

PROPOSITION 15.1. *$L(\underline{m})$ has Seifert form $A' \oplus A''_0 = A^{(m'_0)} \oplus A''_0$, where A', A''_0, A are Seifert forms for $L'(m')$, $L''_0(\underline{m})$, and the knot $L' = (\Sigma', S'_0)$.*

Moreover if $L(\underline{m})$ was fibered, we could take all the above surfaces to be fibers, yielding

PROPOSITION 15.2. *If* $L(\underline{m})$ *is fibered then its algebraic monodromy is* $H'\oplus H''_0 = H^{(m'_0)}\oplus H''_0$, *where* H', H''_0, H *are the monodromies of* $L'(m')$, $L''_0(\underline{m})$ *and* $L' = (\Sigma', S'_0)$.

The main point of the above theorems is that they always can be applied to a knot. A simple induction thus yields the following.

THEOREM 15.3. *Let* L *be a knot and let* L_i, $i = 1, \cdots, r$, *be the results of taking the multilink splice components of* L *and deleting all link components of multiplicity* 0. *Each* L_i *then has at most one link component (so it is an empty link or a* "multiknot"*). A Seifert form for* L *is the sum of Seifert forms of these multiknots. The same holds for algebraic monodromy in the fibered case.*

In view of this theorem, to compute the Seifert forms, and monodromy in the fibered case, of *any* graph knot, it is sufficient to do so for all Seifert knots. In fact, by using non-minimal splice decompositions, it suffices to do it for all Seifert knots given by splice diagrams of the form:

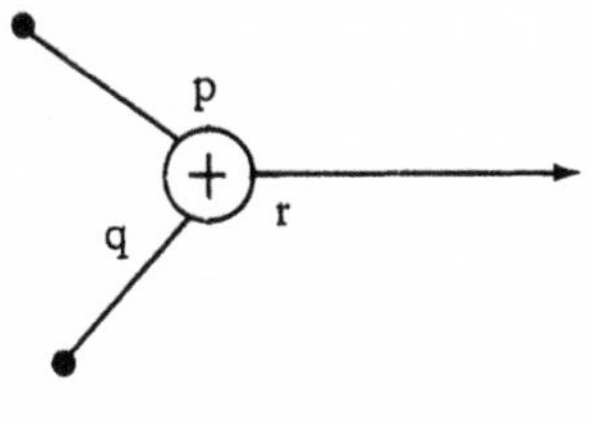

We describe how to do this in the following. A special case of this discussion is the well-known behavior of Seifert form under cabling of knots (see for instance [A'C]).

Seifert knots are always fibered, so we can take a fiber F as our Seifert surface. A, H, and S will denote respectively the Seifert form, monodromy, and intersection form on $H_1(F)$, thought of as matrices with respect to some chosen basis of $H_1(F)$. They satisfy

(15.4) $$H = A^{-1}A^t, \quad S = A - A^t;$$

(15.5) $$A = S(I-H)^{-1}.$$

Thus A determines H and S, and vice versa. (15.5) is an elementary consequence of (15.4), which itself is proved by a simple geometric argument. See for instance Durfee [Du].

(15.6) *If* H *is the monodromy for*

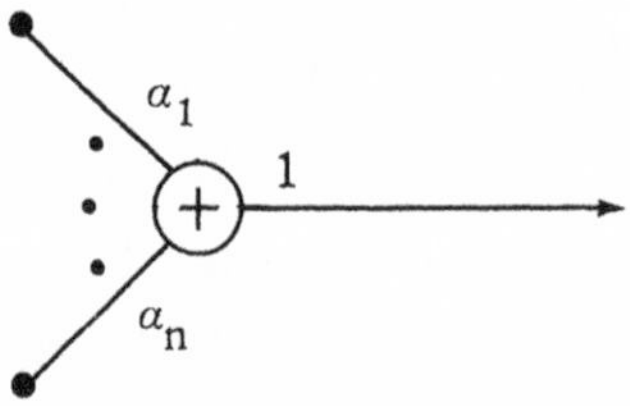

then H^r *is the monodromy for*

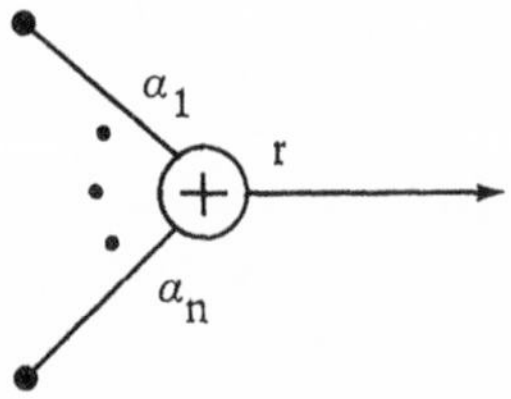

Thus if A *is the Seifert form for the first knot then* $A' = S(I-H^r)^{-1} = (A-A^t)(I-(A^{-1}A^t)^r)^{-1}$ *is the Seifert form for the second.*

We sketch a proof. Let $h: F \to F$ be the monodromy for the first knot, so h has order $\alpha = \alpha_1 \cdots \alpha_n$. A direct homological computation shows that, for r prime to α, the mapping torus of $h^r : F \to F$ is a homology circle, so h^r is still the monodromy of a fibered knot. Since h^r has finite order, this fibered knot is a Seifert knot. The singular fibers still have degrees $\alpha_1, \cdots, \alpha_n$, so it is given by a graph

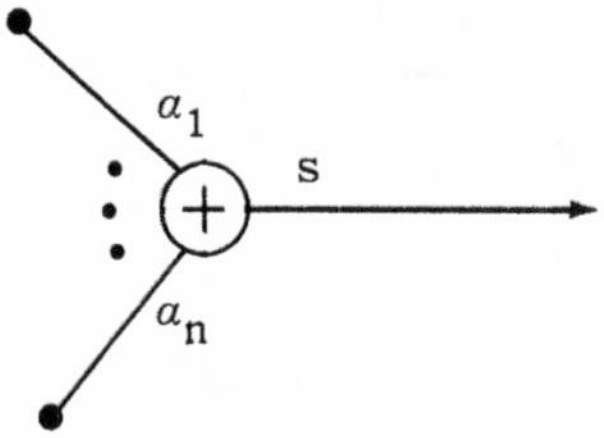

for some s. Comparison with Theorem 13.5 shows $s \equiv r \pmod{\alpha}$, completing the proof.

(15.7) *The Seifert form for the torus knot*

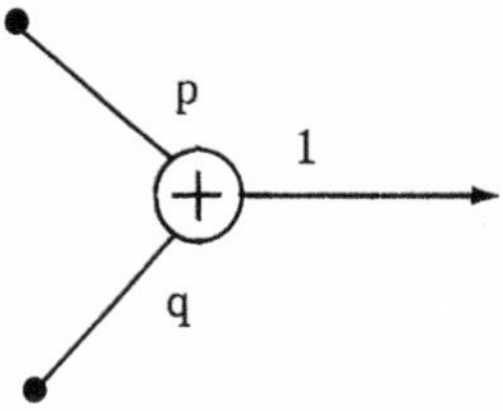

is $-\Lambda_p \otimes \Lambda_q$, *where* Λ_a *is the* $(|a|-1) \times (|a|-1)$ *matrix*

$$\Lambda_a = \operatorname{sign}(a) \begin{pmatrix} 1 & -1 & 0 & \cdots & 0 \\ 0 & 1 & -1 & \cdots & 0 \\ \vdots & & \ddots & \ddots & \\ \vdots & & & 1 & -1 \\ 0 & \cdots & \cdots & 0 & 1 \end{pmatrix}$$

This is well known for p and q positive, see for instance [K-N, §6]. In general it then follows from the fact that reversing the ambient sphere orientation of a knot replaces A by $-A$ (reversing the knot orientation replaces A by A^t).

We illustrate the above with an explicit example. We shall compute the Seifert form for the (2,3)-cable on the degree 3 exceptional fiber in the Poincare sphere:

L:

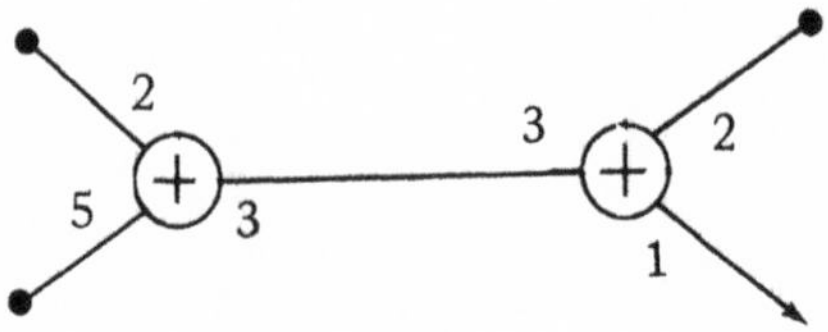

This is the splice of

L′(2):

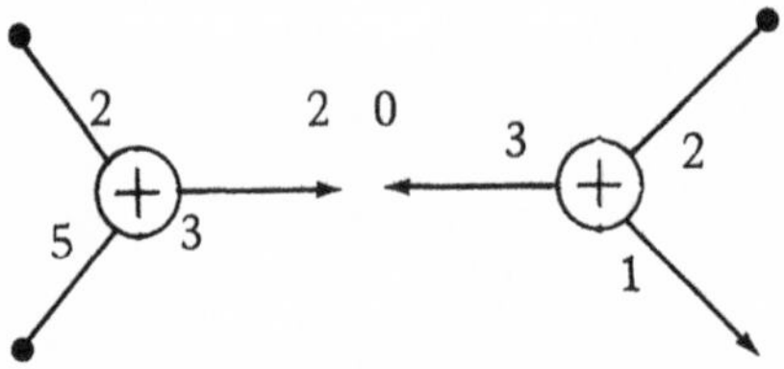

: L″(0,1) ,

so the desired Seifert form is the sum of the Seifert forms of L′(2) and

L''_0:

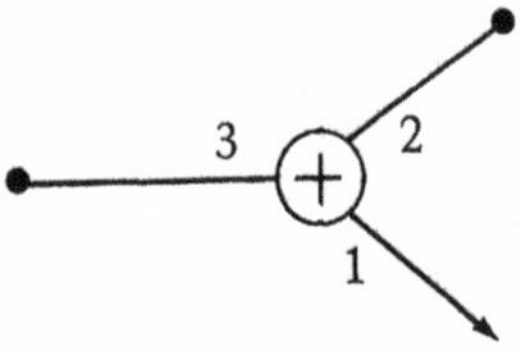

The (2,5)-torus knot

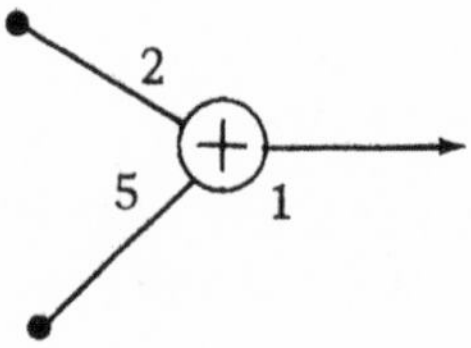

has Seifert form

$$\begin{pmatrix} -1 & 1 & 0 & 0 \\ 0 & -1 & 1 & 0 \\ 0 & 0 & -1 & 1 \\ 0 & 0 & 0 & -1 \end{pmatrix}$$

by (15.7). Thus by (15.6) one computes

$$A(L') = \begin{pmatrix} 1 & 1 & -1 & -1 \\ 0 & 1 & 1 & -1 \\ -1 & 0 & 1 & 1 \\ -1 & -1 & 0 & 1 \end{pmatrix} .$$

Now

$$A(L'(2)) = \begin{pmatrix} A(L') & A(L')^t \\ A(L') & A(L') \end{pmatrix} ,$$

so

$$A(L) = \begin{pmatrix} A(L') & A(L')^t & 0 \\ A(L') & A(L') & 0 \\ 0 & 0 & A(L''_0) \end{pmatrix}$$

where

$$A(L''_0) = \begin{pmatrix} -1 & 1 \\ 0 & -1 \end{pmatrix} ,$$

is the Seifert form of the trefoil.

Theorem 15.3 enables one to give many examples of graph knots with identical Seifert forms. The simplest such examples are the cabled torus knots

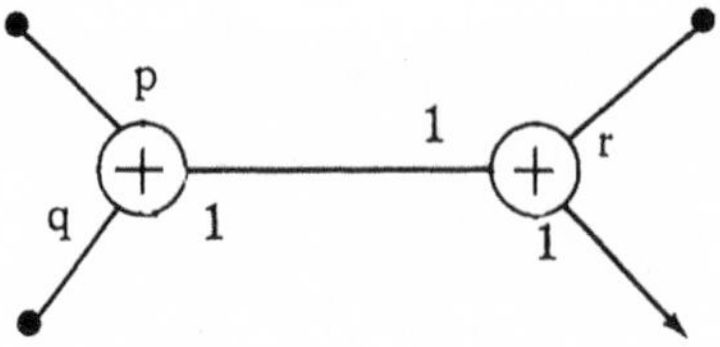

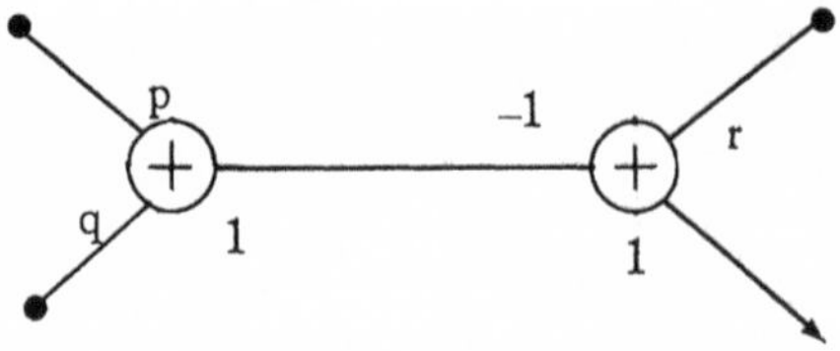

Indeed, the (r,1)-cable and the (r−1)-cable of any knot have identical Seifert forms. More generally, the (r,s)-cable on a knot L has the same Seifert form as the sum of the (r,1)-cable on L and an (r,s)-torus knot.

A slightly more interesting example is the following. Consider the following four cabled torus knots.

L_1 :

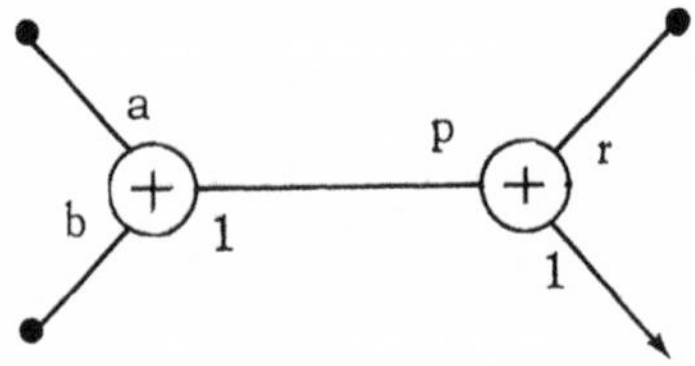

L_2 :

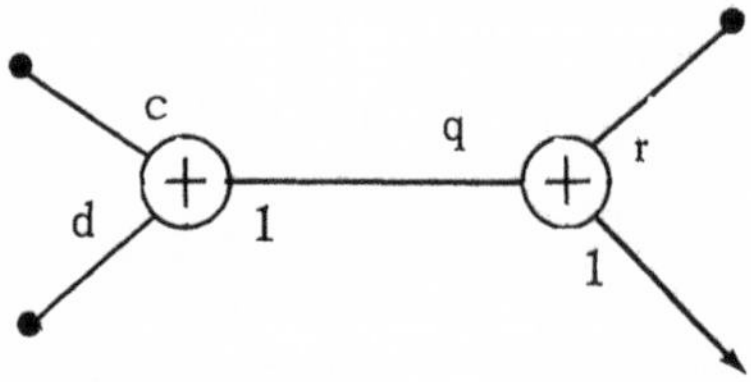

L_3 :

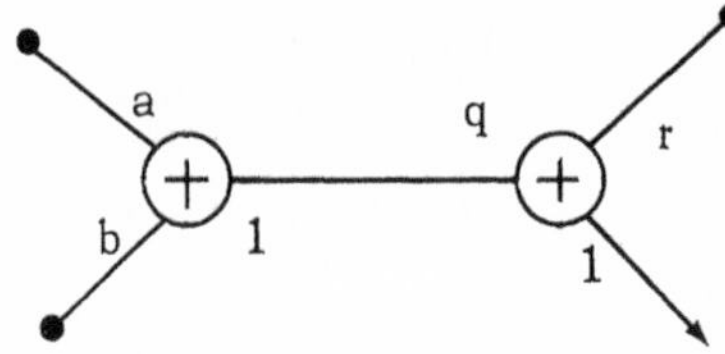

L_4 :

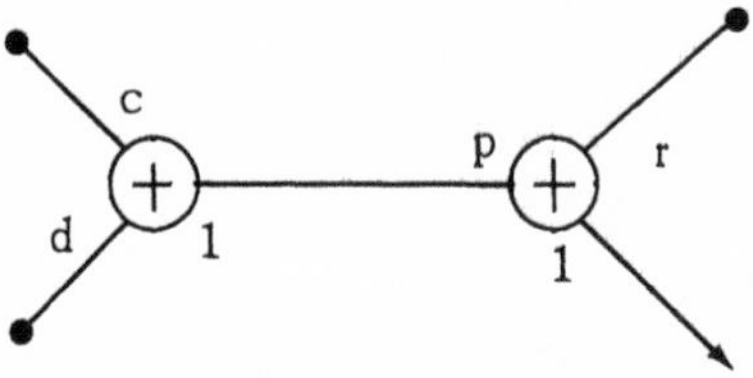

Then $L_1 \# L_2$ has the same Seifert form as $L_3 \# L_4$. Now, if p and q both exceed abr and cdr, then all four knots are algebraic (that is, links of plane curve singularities). This gives then a relation in algebraic bordism of algebraic knots which answers a question of Rudolph, see Litherland [Li], who showed that an algebraic knot cannot be algebraically bordant to a sum of other algebraic knots.

Chapter IV

EXAMPLES

16. *Trivial geometric monodromy*

If a fibered link or multilink has finite order geometric monodromy then it must be a Seifert multilink. It is thus a simple but amusing exercise to determine when the geometric monodromy can actually be trivial. For links in S^3 there are four infinite classes of examples (up to reversal of orientation) (edge weights equal to 1 are omitted):

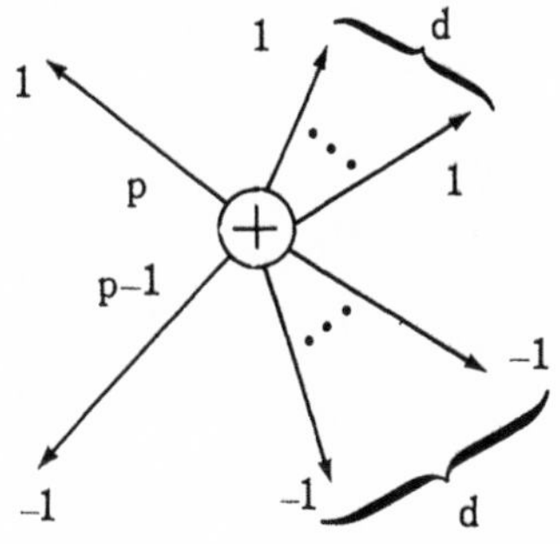

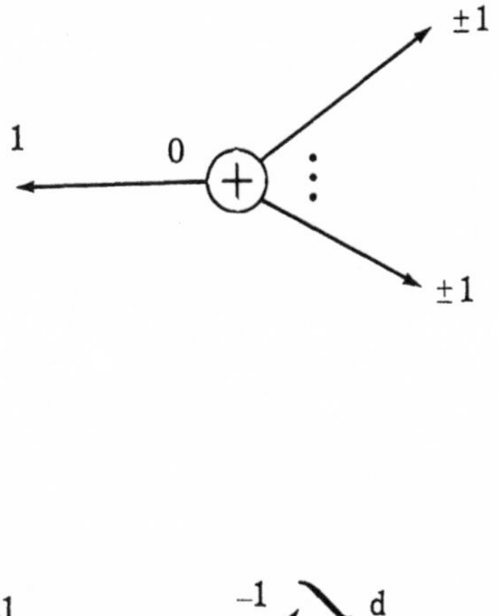

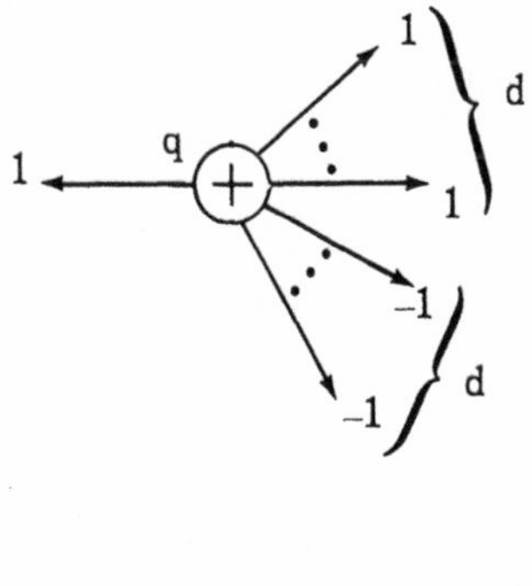

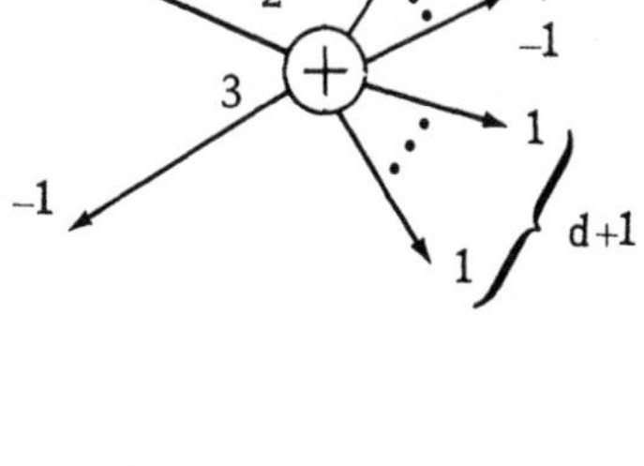

The last example for $d = 0$ can be described as the link of the singularity at the origin of the real analytic surface $V = F^{-1}(0) \subset \mathbb{C}^2$ where

$$f : \mathbb{C}^2 \to \mathbb{C} \quad \text{is} \quad f(x,y) = \overline{x}(x^3 + y^2)\overline{y}\ .$$

This is "weighted homogeneous": The $\mathbb{C}^*$-action $t(x,y) = (t^2x, t^3y)$ satisfies $f(t(x,y)) = f(x,y)t\,|t|^{10}$. The fact that f has "degree 1" for this $\mathbb{C}^*$-action is just a translation of the fact of trivial geometric monodromy; one sees the triviality of monodromy very explicitly via:

$$f^{-1}(1) \times \mathbb{C}^* \xrightarrow{\cong} \mathbb{C}^2 - V$$

$$((x,y),t) \longmapsto t(x,y) = (t^2x, t^3y)\ .$$

It is amusing to draw a picture of a fiber of this link in S^3. First the link itself:

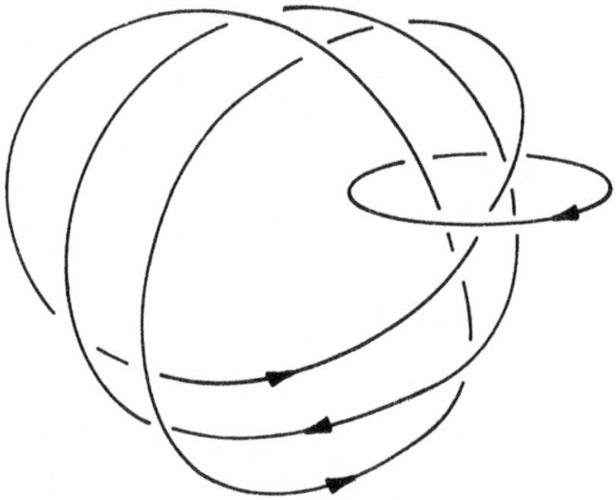

Now the same link isotopically deformed to make a fiber evident:

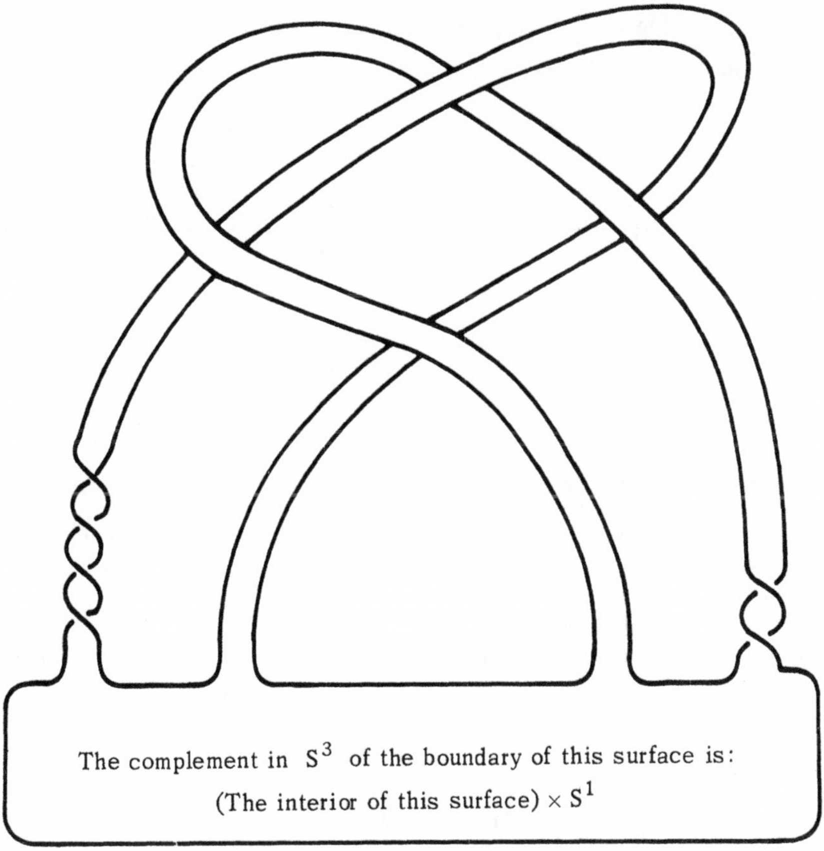
The complement in S^3 of the boundary of this surface is:
(The interior of this surface) $\times S^1$

17. *Further examples*

We record in this section some computations relevant to several "obvious" examples; we leave the (easy) verifications mostly to the reader.

a) *Iterated torus knots.* (See [S-W] for a different treatment.)

An iterated torus knot (S^3, K) is, by definition, a (p_n, q_n)-cable on a (p_{n-1}, q_{n-1})-cable on a (p_1, q_1)-cable on the unknot, for some pairs (p_i, q_i) and thus has diagram

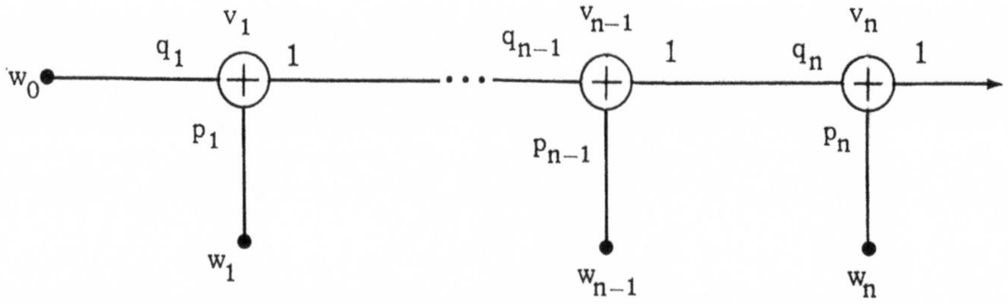

where we have given the non-arrow head vertices labels for reference. The diagram is in normal form if and only if no p_i is 0 or 1 (and thus no q_i is 0). The knot is algebraic if and only if all p_i and q_i are positive, and $q_k > p_k p_{k-1} q_{k-1}$ for $k = 2, \cdots, n$; its Newton pairs are then $(p_1, q_1), (p_2, (q_2 - p_2 p_1 q_1)), \cdots, (p_n, (q_n - p_n p_{n-1} q_{n-1}))$. The linking numbers of the knot with nonsingular fibers corresponding to the non-arrowhead vertices are

$$\underline{m}(v_i) = q_i p_i \cdots p_n$$

$$\underline{m}(w_0) = p_1 \cdots p_n$$

$$\underline{m}(w_i) = q_i p_{i+1} \cdots p_n .$$

If the diagram is in normal form, so that no p_i is 0 or 1 , then of course no q_i is 0, so we see that the knot fibers, and the genus of the fiber is

$$\sum_1^n |\underline{m}(v_i)| - \sum_0^n |\underline{m}(w_i)| .$$

If we assume that all p_i, q_i are positive, this becomes

$$-q_n + p_n(q_n - q_{n-1} + p_{n-1}(q_{n-1} - q_{n-2} + p_{n-2}(\cdots(q_2 - q_1 + p_1(q_1 - 1)\cdots).$$

The characteristic polynomial of the monodromy action (which is the same as the Alexander polynomial) is given by

$$(t-1)\frac{\prod_1^n (t^{|\underline{m}(v_i)|}-1}{\prod_0^n (t^{|\underline{m}(w_i)|}-1)}\,.$$

The algebraic monodromy has finite order (as for any solvable knot).

The multilink splice components of (S^3,K) are

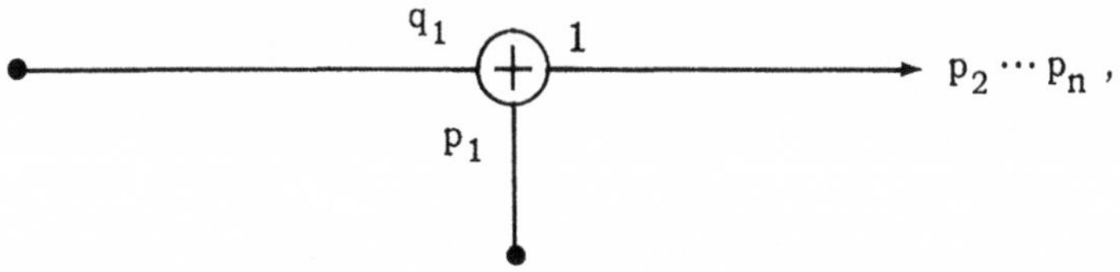

and

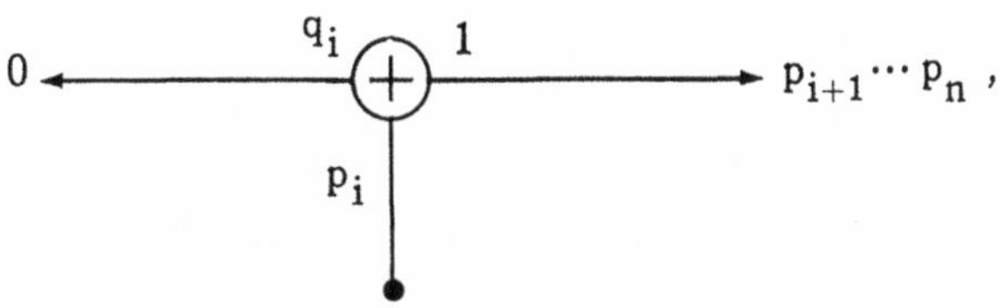

for $i = 2, \cdots, n$.

Thus the Seifert form and algebraic monodromy of (S^3, K) are the sums of those corresponding to the multi-knots

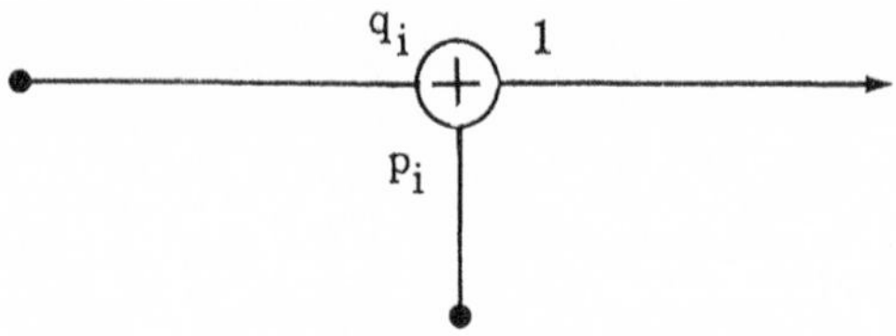

$(i=1,\cdots,n)$,

which in turn are given by 15.7, 15.4, and the discussion at the beginning of section 15. (See also the paper [A'C] of A'Campo.)

b) *Coaxial torus links.* These are solvable links $L = (S^3, K_1 \cdots K_n)$ where the K_i are (p_i, q_i)-torus knots tied on a series of progressively smaller tori (for beautiful pictures see [A, p. 490 ff] or [B-K, p. 603]). They correspond to splice diagrams of the form

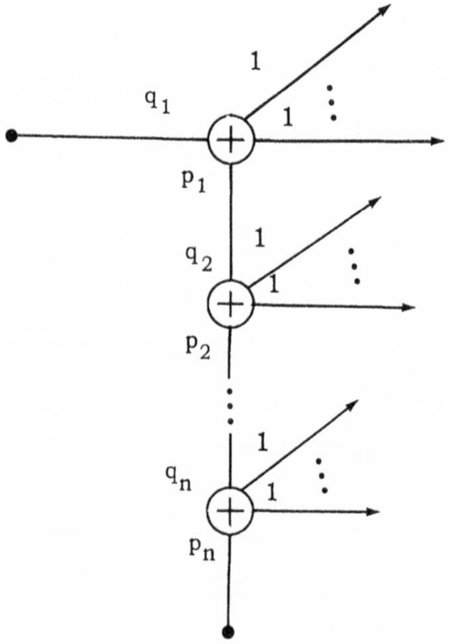

,

which is minimal as long as p_n and q_1 are not 0 or 1 and $p_i/q_i \neq p_{i+1}/q_{i+1}$ for $i = 1, \cdots, n-1$. Assuming that the diagram is minimal, L is algebraic iff $p_i/q_i > p_{i+1}/q_{i+1}$ for $i = 1, \cdots, n-1$.

Using our methods it is easy to recover the examples due to Marie-Claire Grima of pairs of fibered coaxial torus-knots having, for example, the same characteristic polynomial of the monodromy. The following links, for example, are both fibered, with finite-order algebraic monodromy, and the same characteristic polynomial:

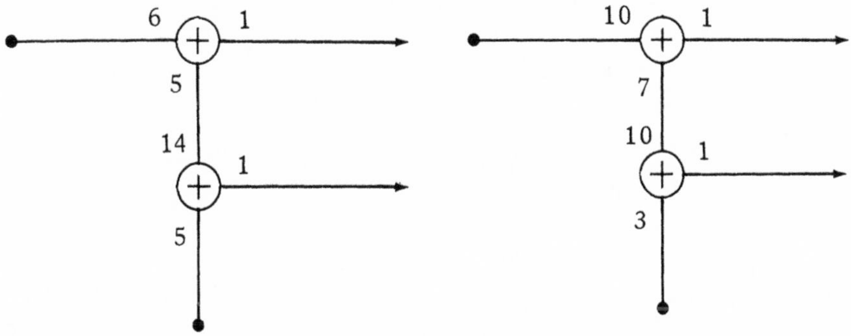

The characteristic polynomial is easily seen to be

$$(t-1)\,\frac{(t^{60}-1)(t^{100}-1)}{(t^{10}-1)(t^{20}-1)}\,,$$

a polynomial of degree 131.

c) *Square and granny knots*

We include the diagrams of these simply for amusement, and leave the computations of the genus, Alexander polynomials, and Seifert forms as an easy exercise:

The square knot is the connected sum of a trefoil with its mirror image

and thus has diagram

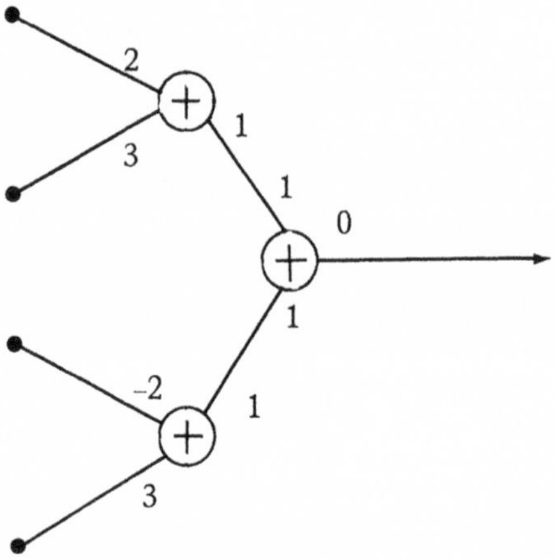

and multilink splice decomposition

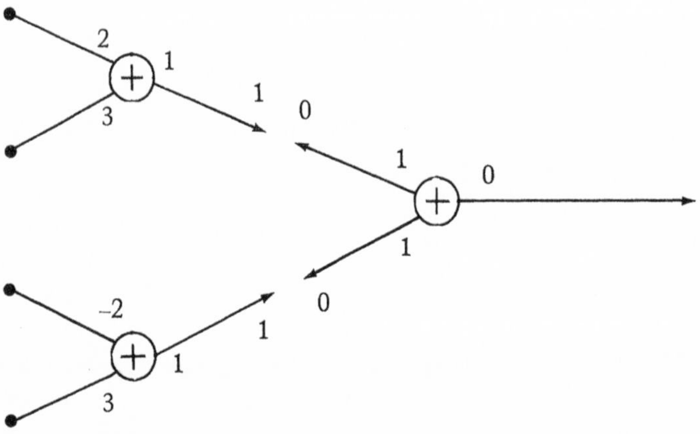

The "Granny" knot is the connected sum of two identical trefoils, and thus has diagrams as above, but with the -2 replaced by $+2$:

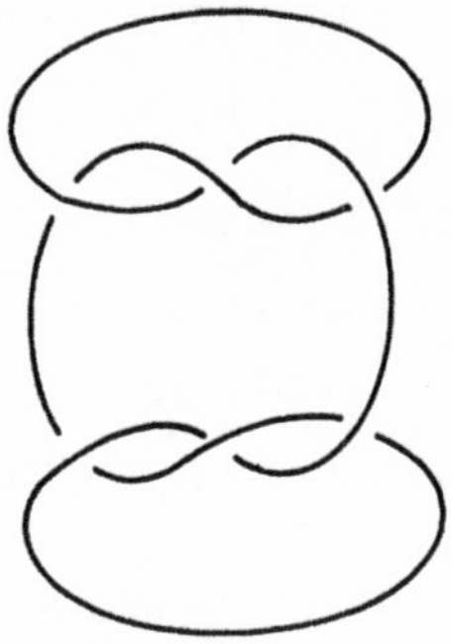

d) *Tree-Chains.* This class of links is well exemplified by the following sample:

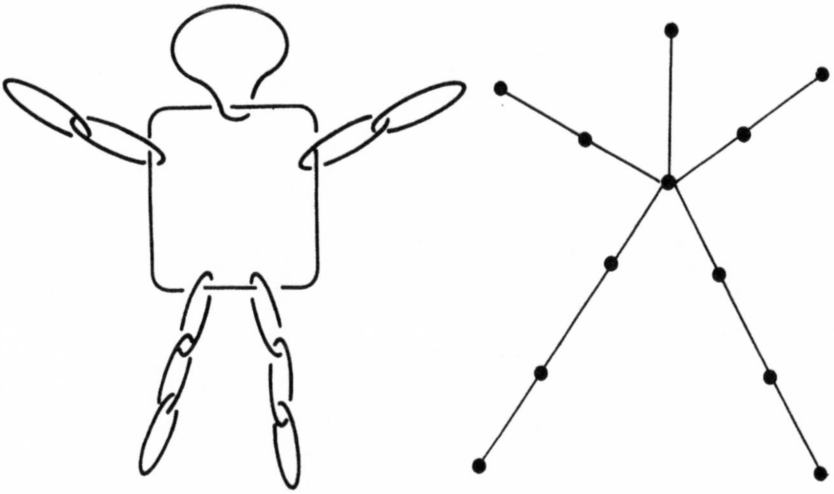

More generally, given any tree T, the corresponding tree-chain in S^3 has as components one unknotted oriented circle for each node of the tree, chosen so that 2 circles link, with linking number 1, iff the two nodes are joined. The main points are:

PROPOSITION 17.1. 1) *Tree chains are solvable links.*

2) *A tree chain with* n *components which is not split is fiberable with fiber the* n*-punctured sphere and trivial algebraic monodromy.*

Proof. 1) The reader may easily check that tree-chains correspond precisely to those diagrams for which every node has an arrowhead with a 0 weight emerging from it, thus:

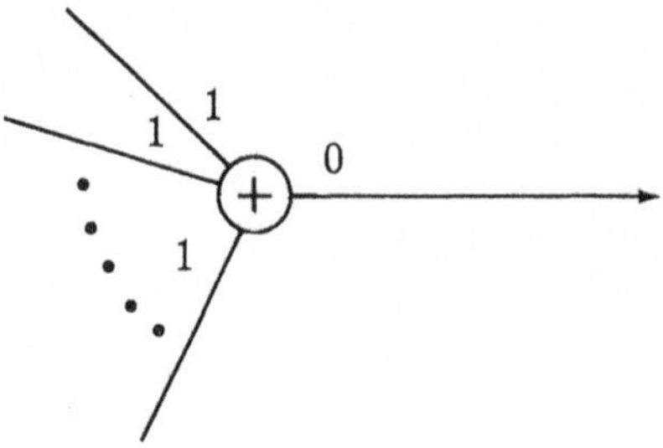

2) If v is a node as in 1), then $\underline{m}(S_v) = 1$, so the link is indeed fibered. Since a normalized graph for the tree-chain will obviously have no vertices of the form

,

we see that the characteristic polynomial of the monodromy is

$$(t-1)\prod_{v}(t-1)^{(\text{valence}(v)-2)} ,$$

the product being taken over the nodes

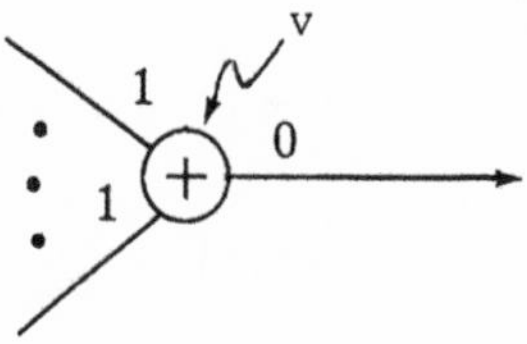

of the associated diagram in normal form. An easy induction shows that this is $(t-1)^{n-1}$. It follows that, if F is the fiber, then F has n boundary components and rank $H_1(F) = n-1$; since F is connected, F must be a sphere with n punctures. One now sees directly by geometry, or via Corollary 11.5, that the algebraic monodromy is trivial.

Chapter V
RELATION TO PLUMBING

18. *Plumbing*

To fix notation we recall briefly the basic concepts of plumbing. The following section 19 describes plumbing arising from resolution of singularities. Since we are mainly interested in homology spheres, only plumbing bundles over spheres according to a tree is of interest to us. For our purposes, therefore, a *plumbing tree* is a finite contractible 1-complex Δ with an integer weight e_i assigned to each vertex i.

The *plumbed 3-manifold* $M(\Delta)$ associated with Δ is constructed as follows: Let d_i be the degree of vertex i (number of incident edges). Let F_i be a genus zero surface with d_i boundary components (S^2 with d_i open discs removed) and let $E_i \to F_i$ be an oriented circle bundle over F_i with a chosen trivialization of $E_i|\partial F_i$ and with euler number e_i (this is the cross-section obstruction, and is well defined once the trivialization $E_i|\partial F_i$ is chosen). $M(\Delta)$ is pasted together from the E_i as follows: whenever vertices i and j are connected by an edge of Δ, we paste a boundary component $S^1 \times S^1$ of E_i to a boundary component $S^1 \times S^1$ of E_j by the map $(x,y) \to (y,x)$ which exchanges base and fiber coordinates.

It is convenient also to allow disconnected plumbing graphs, with the convention that if $\Delta = \Delta_1 + \Delta_2$ (disjoint union) then $M(\Delta) = M(\Delta_1) \# M(\Delta_2)$.

More generally a *decorated plumbing tree* (or *plumbing diagram*) Δ is like a plumbing tree except that some vertices of degree 1 in Δ are left unweighted, and are drawn in Δ as arrowheads, for example

$\Delta =$ 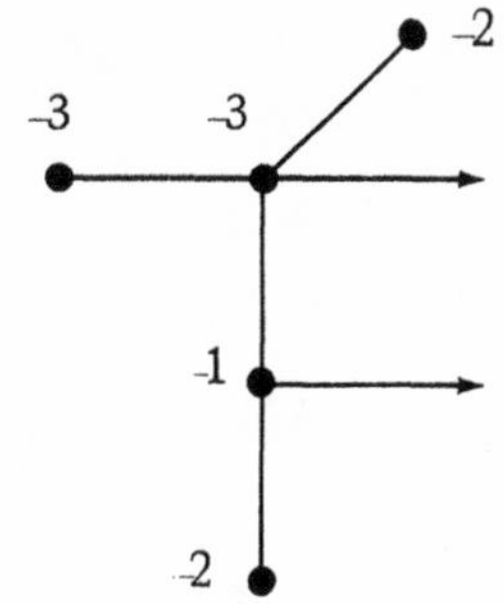

Such a decorated plumbing tree represents a link in a 3-manifold, $K(\Delta) \subset M(\Delta)$, as follows (see [Ne 3, Appendix, part 2]). Let Δ_0 be Δ with arrows deleted and put $M(\Delta) = M(\Delta_0)$. Each arrow j of Δ is attached at some vertex i of Δ_0, and to this arrow we associate a fiber S_j of the bundle E_i used in plumbing. We take the S_j for different arrows j to be disjoint; $K(\Delta)$ is the union of these S_j.

For the purpose of this definition we need Δ_0 nonempty, so if $\Delta = \leftrightarrow$ or $\updownarrow$, we replace it by $\longleftarrow\overset{\pm 1}{\bullet}\longrightarrow$ or $\longleftarrow\overset{\pm 1}{\bullet}$ respectively.

If $M(\Delta)$ is a homology sphere then $(M(\Delta), K(\Delta))$ is a graph link. Later (Section A4) we describe how to construct a plumbing representation for any given graph link. For the moment however we do not need to restrict to graph links, that is, $M(\Delta)$ need not be a homology sphere.

The exterior $M_0(\Delta) = M(\Delta) - \operatorname{int} N(K(\Delta))$, where $N(K(\Delta))$ is a tubular neighborhood of $K(\Delta)$, can be constructed directly by "plumbing." To each non-arrowhead vertex i of degree d_i in Δ take an S^1-bundle $E_i \to F_i$ with d_i boundary components as before (so a trivialization of $\partial E_i \to \partial F_i$ is chosen, and e_i is the cross section obstruction); paste as before. Since edges ending in an arrowhead now no longer correspond to a pasting, they give boundary components of the resulting manifold $M_0(\Delta)$. $M(\Delta)$ can be recovered by capping off these boundary components with solid tori.

Some additional notation will be helpful. Let $\mathcal{V} = \mathcal{V}(\Delta)$ be the set of vertices of Δ, and $\mathcal{A} = \mathcal{A}(\Delta)$, $\overline{\mathcal{A}} = \overline{\mathcal{A}}(\Delta)$ the subsets of arrowheads and non-arrowheads respectively, so $\mathcal{A} \cup \overline{\mathcal{A}} = \mathcal{V}$. For any $i \in \mathcal{V}$, let $\mathcal{V}(i)$, $\mathcal{A}(i)$, $\overline{\mathcal{A}}(i)$ be the set of vertices in $\mathcal{V}$, $\mathcal{A}$, or $\overline{\mathcal{A}}$ which are adjacent to i in Δ, that is, connected to i by an edge.

Let $A(\Delta)$ be the matrix

$$A(\Delta) = (A_{ij})_{i,j \in \overline{\mathcal{A}}(\Delta)}$$

$$A_{ij} = \begin{cases} e_i & i = j \\ 1 & j \in \overline{\mathcal{A}}(i) \\ 0 & j \notin \overline{\mathcal{A}}(i) \cup \{i\} . \end{cases}$$

$A(\Delta)$ may be familiar to the reader as the intersection form of the 4-manifold, obtained by plumbing disc bundles according to Δ_0, of which $M(\Delta)$ is the boundary. See for instance [H-N-K], where the following lemma is deduced. Think of $A(\Delta)$ as the matrix of a map $Z^{\overline{\mathcal{A}}(\Delta)} \to Z^{\overline{\mathcal{A}}(\Delta)}$.

LEMMA 18.1. $H_1(M(\Delta)) = \operatorname{cok}(A(\Delta))$, *so* $M(\Delta)$ *is a homology sphere if and only if* $\det(A(\Delta)) = \pm 1$.

For an arrowhead $j \in \mathcal{A}(\Delta)$, let M_j be an oriented meridian in $\partial M_0(\Delta)$ of the corresponding link component $S_j \subset K(\Delta)$. For $i \in \overline{\mathcal{A}}(\Delta)$, let $S_i \subset M_0(\Delta)$ be an oriented fiber of the bundle E_i used in plumbing.

As in Section 3, a cohomology class $\underline{m} \in H^1(M_0(\Delta))$ defines a "multilink structure" on $(M(\Delta), K(\Delta))$. The *multiplicity* of the j-th component S_j of $K(\Delta)$ with respect to $\underline{m}$ is

$$m_j = \underline{m}(M_j), \quad j \in \mathcal{A}(\Delta) .$$

For $i \in \overline{\mathcal{A}}(\Delta)$, the number

$$\ell_i = \underline{m}(S_i) ,$$

which we call the *multiplicity of vertex* i, can be considered as the linking number of S_i with the multilink. This agrees with our earlier concepts if $M(\Delta)$ is a homology sphere.

For $i \in \overline{\mathfrak{A}}(\Delta)$, let

$$s_i = \sum_{j \in \mathfrak{A}(i)} m_j ,$$

so s_i is the total multiplicity of all link components at vertex i.

THEOREM 18.2. *Let* $\underline{m} \in H^1(M_0(\Delta))$ *and let notation be as above. Then for any* $i \in \overline{\mathfrak{A}}(\Delta)$

$$\sum_{k \in \overline{\mathfrak{A}}(\Delta)} A_{ik} \ell_k + s_i = 0 .$$

Equivalently, if $\vec{\ell}$, $\vec{s}$ *are the column vectors* $(\ell_i)_{i \in N}$ *and* $(s_i)_{i \in N}$, *then*

$$A(\Delta) \vec{\ell} + \vec{s} = 0 .$$

Conversely, any set of ℓ_i *and* m_j *which solve this equation is the set of multiplicities for some* $\underline{m} \in H^1(M_0(\Delta))$.

REMARK. It follows that, if $M(\Delta)$ is not a homology sphere, in general not every collection of individual link component multiplicities $\{m_j, j \in \mathfrak{A}(\Delta)\}$ can be realized, since $A(\Delta)$ will not be invertible. Moreover, if $M(\Delta)$ is not even a rational homology sphere, that is $\det(A(\Delta)) = 0$, then the m_j do not determine the ℓ_i.

Proof. Let E_i be the bundle corresponding to vertex $i \in \overline{\mathfrak{A}}(\Delta)$ used in the plumbing construction of $M_0(\Delta)$. For each $j \in \mathfrak{O}(i)$, the trivialization of the corresponding boundary component $(\partial E_i)_j$ of E_i determines a section $M_{ij} \subset (\partial E_i)_j$ up to isotopy. The fact that e_i is the cross section obstruction can be written as the homology relation

$$\sum_{j \in \mathfrak{O}(i)} M_{ij} + e_i S_i = 0 \text{ in } H_1(E_i) .$$

But now $M_{ij} = M_j$ if $j \in \mathfrak{A}(i)$ and M_{ij} is isotopic in $M_0(\Delta)$ to S_j if $j \in \overline{\mathfrak{A}}(i)$. Thus this homology relation can be written

$$(*) \qquad \sum_{k \in \bar{\mathfrak{A}}} A_{ik} S_k + \sum_{j \in \mathfrak{A}(i)} M_j = 0 \text{ in } H_1(M_0(\Delta)) .$$

Applying $\underline{m}$ to this equation gives the equation to be proved.

The converse realizability statement would follow if we knew that the relations (*) for $i \in \bar{\mathfrak{A}}(\Delta)$ give a presentation for $H_1(M_0(\Delta))$ in terms of the generators S_i and M_j. This is indeed true, and follows directly by computing $H_1(M_0(\Delta))$ by the Mayer Vietoris sequence. We omit the details; note that if $\mathfrak{A}(\Delta) = \phi$ it is a restatement of Lemma 18.1.

Example. If we take the plumbing graph

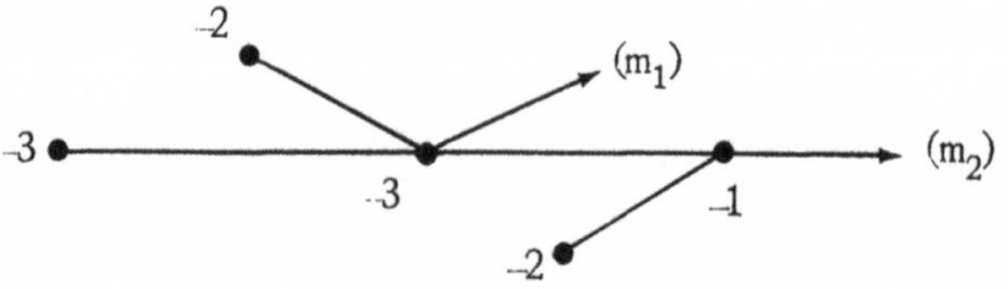

of our earlier example then, numbering non-arrowhead vertices from left to right,

$$A(\Delta) = \begin{pmatrix} -3 & 0 & 1 & 0 & 0 \\ 0 & -2 & 1 & 0 & 0 \\ 1 & 1 & -3 & 0 & 1 \\ 0 & 0 & 0 & -2 & 1 \\ 0 & 0 & 1 & 1 & -1 \end{pmatrix} .$$

One computes that $\det(A(\Delta) = -1$, so this plumbing gives a graph link. Moreover

$$-A(\Delta)^{-1} = \begin{pmatrix} 1 & 1 & 2 & 2 & 4 \\ 1 & 2 & 3 & 3 & 6 \\ 2 & 3 & 6 & 6 & 12 \\ 2 & 3 & 6 & 7 & 13 \\ 4 & 6 & 12 & 13 & 26 \end{pmatrix}$$

so

$$\vec{\ell} = -A(\Delta)^{-1}\begin{pmatrix} 0 \\ 0 \\ m_1 \\ 0 \\ m_2 \end{pmatrix} = \begin{pmatrix} 2m_1 + 4m_2 \\ 3m_1 + 6m_2 \\ 6m_1 + 12m_2 \\ 6m_1 + 13m_2 \\ 12m_1 + 26m_2 \end{pmatrix}.$$

Thus the most general multilink on Δ has multiplicities

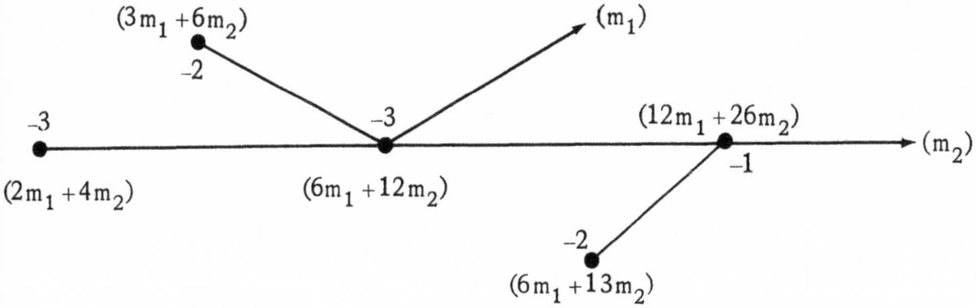

In fact this link is a link in S^3 and it is algebraic (that is, the link of a plane curve singularity). The reason is that the plumbing graph Δ_0 (Δ with arrows deleted) can be reduced to the trivial graph by "blowing down" (-1)-vertices. We see this example again from an explicitly algebraic point of view in the next section.

We remind the reader of some operations on plumbing trees that do not change the represented link. We shall include the effect on multiplicities, so let $\Delta(\underline{m})$ denote a plumbing tree Δ plus an assignment of multiplicities as above.

THEOREM 18.3. *The following operations on* $\Delta(\underline{m})$ *do not alter the represented graph multilink. Moreover two graphs* $\Delta(\underline{m})$ *and* $\Delta'(\underline{m}')$ *which represent the same multilinks are related by a sequence of these moves and their inverses.*

1. $\Delta(\underline{m}) \to \Delta(-\underline{m})$
2. (± 1)-*Blowing down*:

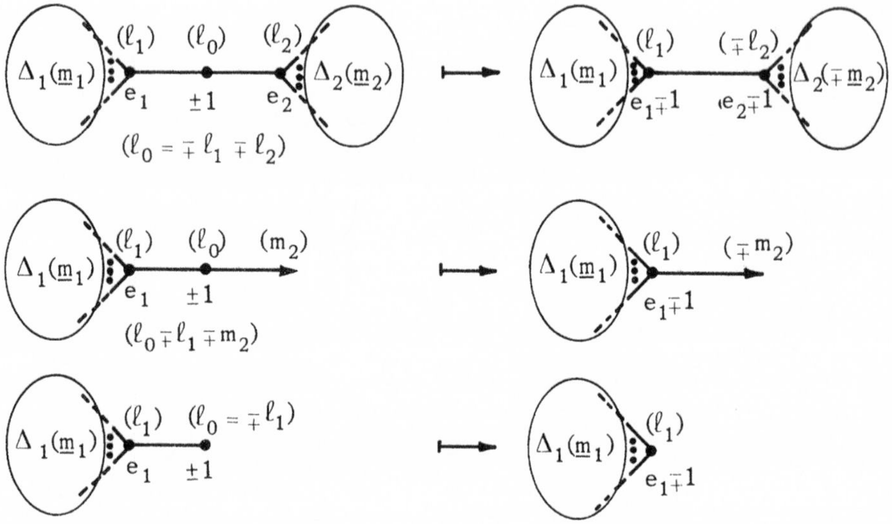

(disjoint union)

3. *0-chain absorption*:

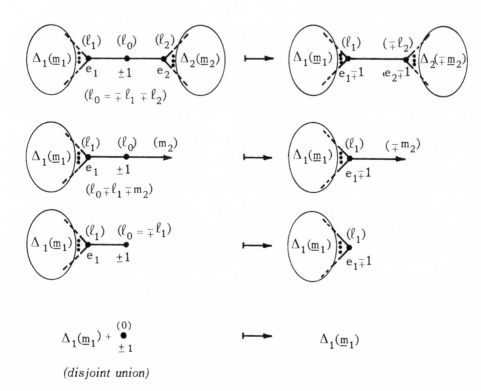

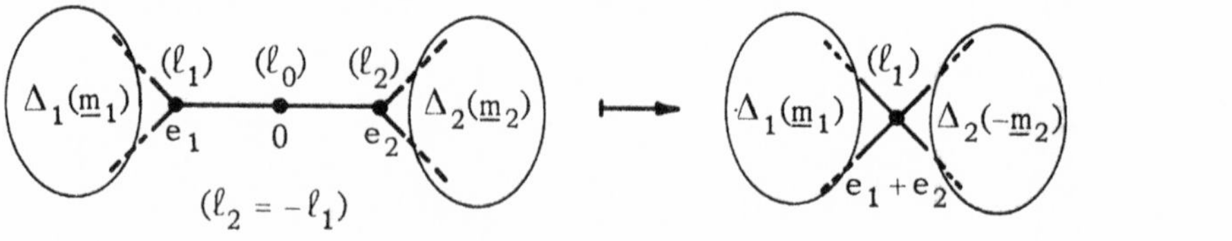

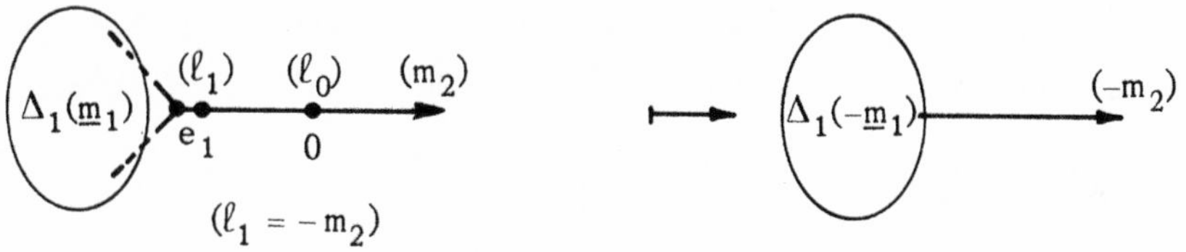

4. *Splitting:*

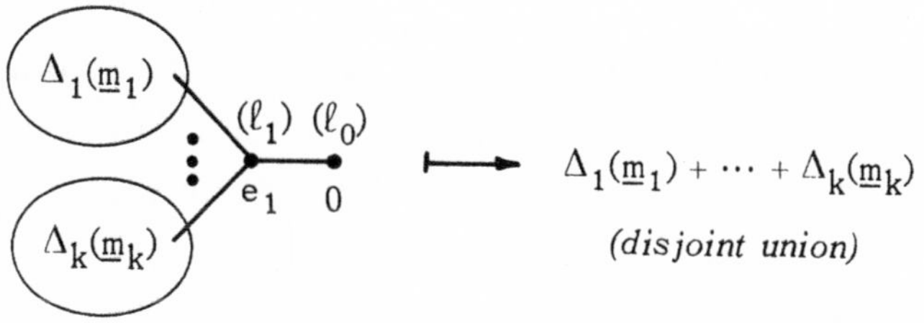

Without the description of the effect on multiplicities, this theorem is proved in [Ne 3, Theorem 3.2 and Appendix]. The information on multiplicities is valid by Theorem 18.2. The theorem is valid even if $M(\Delta)$ is not a homology sphere. The proof in [Ne 3] proceeded by using the moves to reduce a plumbing diagram to a unique normal form. For graph links this normal form is particularly simple, so we describe it.

A vertex i of Δ is called a *node* if it has degree ≥ 3.

THEOREM 18.4. *Up to replacing* $\Delta(\underline{m})$ *by* $\Delta(-\underline{m})$, *any graph link is given by a unique plumbing diagram satisfying:*

$e_i \leq -2$ *for every vertex* i *of* Δ *which is not an arrowhead or node.*

We call such a plumbing diagram the *normal form* plumbing diagram for the graph link. (This theorem is valid even if $M(\Delta)$ is not a homology sphere, but then the present normal form may not be the same one as used in [Ne 3].)

If a plumbing diagram admits no blowing down, zero-chain absorption, or splitting, we call it *minimal.* Clearly normal form diagrams are minimal.

19. *Resolution of plane curve singularities*

This review of resolution of plane curve singularities is not needed for the general discussion but it is important as motivation for much of this chapter.

Consider a singular complex analytic germ $f:(\mathbf{C}^2,0) \to (\mathbf{C},0)$. Its *link* is $L(f) = (S^3_\varepsilon, S^3_\varepsilon \cap f^{-1}(0))$ for a suitably small sphere S^3_ε about $0 \in \mathbf{C}^2$. Since the components of $V = f^{-1}(0)$ have multiplicities, $L(f)$ may be considered as a multilink. We write $L(f) = (S^3, K(f))$.

We shall remind the reader of how a resolution of f yields a plumbing description of $L(f)$. A reference for this is [H-N-K].

Let $U \subset \mathbf{C}^2$ be a small neighborhood of $0 \in \mathbf{C}^2$ on which f is defined. By repeated blowing up over the origin we obtain a complex space $\widetilde{U}$ and analytic map $\pi: \widetilde{U} \to U$ such that if $\widetilde{f} = f \circ \pi$, then:

(i) $\pi|(\widetilde{U} - \pi^{-1}(0)): \widetilde{U} - \pi^{-1}(0) \to U - \{0\}$ is a biholomorphic isomorphism.

(ii) The *exceptional divisor* $\pi^{-1}(0) = D = D_1 \cup \cdots \cup D_n$ is a union of curves $D_i \cong \mathbf{CP}_1$ meeting each other transversely.

(iii) $\widetilde{f}^{-1}(0) = D \cup \widetilde{V}$ where $\widetilde{V} = \widetilde{V}_1 \cup \cdots \cup \widetilde{V}_k$ is a disjoint union of nonsingular curves meeting D transversely ($\widetilde{V}$ is called the *proper transform* of V). Each $\widetilde{V}_i$ meets D in just one point (this is achieved by choosing the neighborhood U small enough)

(iv) If C is one of the D_i or $\widetilde{V}_j$, then for each point $p \in C$ there exist holomorphic coordinates (u,v) about p in which $C = \{u=0\}$ and $\widetilde{f}(u,v) = u^\ell v^m$. Then ℓ is called the *multiplicity* of C as a component of $\widetilde{f}^{-1}(0)$.

Note that (iv) implies much of ii) and iii), namely that $\widetilde{f}^{-1}(0)$ is nonsingular except for normal crossings.

We see $S^3_\varepsilon \cong \pi^{-1}(S^3_\varepsilon)$ as the boundary of a tubular neighborhood of D. This represents it as the boundary of the 4-manifold obtained by plumbing together the normal bundles of the D_i according to the intersection configuration of the D_i. Moreover $K(f) \cong \pi^{-1}(S^3_\varepsilon \cap V)$ is then the union of some circle fibers of the boundaries of these disc bundles.

We illustrate this by computing an example. Let

$$f(x,y) = (x^6-y^4)^2 + x^5y(x^3+y^2)^2 .$$

Let $U = \mathbf{C}^2$ and let $\widetilde{U}^1$ be U blown up once at the origin. Then $\widetilde{U}_1$ can be covered by two charts with coordinates (x_1,y_1) and $(\overline{x}_1,\overline{y}_1)$, related by $x_1 = \overline{x}_1\overline{y}_1$, $y_1 = 1/\overline{x}_1$ on their overlap $\overline{x}_1 \neq 0$, $y_1 \neq 0$. With respect to these charts the projection map $\pi_1 : \widetilde{U}_1 \to U$ is given by $\pi(x_1,y_1) = (x_1,x_1y_1)$, $\pi(\overline{x}_1,\overline{y}_1) = (\overline{x}_1\overline{y}_1,\overline{y}_1)$, and the exceptional divisor D_1 is given by $x_1 = 0$ in the one chart, $\overline{y}_1 = 0$ in the other. We can write $\widetilde{f}_1 = f \circ \pi_1$ in these coordinate systems as

$$\begin{aligned}\widetilde{f}_1(x_1,y_1) &= (x_1^6-(x_1y_1)^4)^2 + x_1^5(x_1y_1)(x_1^3+(x_1y_1)^2)^2\\ &= x_1^8[(x_1^2-y_1^4)^2 + x_1^2y_1(x_1+y_1^2)^2]\\ \widetilde{f}_1(\overline{x}_1,\overline{y}_1) &= ((\overline{x}_1\overline{y}_1)^6-\overline{y}_1^4)^2 + (\overline{x}_1\overline{y}_1)^5\,y_1((\overline{x}_1\overline{y}_1)^3+\overline{y}_1^2)^2\\ &= \overline{y}_1^8[(\overline{x}_1^6\overline{y}_1^2-1)^2 + \overline{x}_1^5\overline{y}_1^2(\overline{x}_1^3\overline{y}_1+1)^2] .\end{aligned}$$

Thus $\widetilde{f}_1^{-1}(0) = 8D_1 \cup V_1$, where $V_1 = \{(x_1,y_1) | (x_1^2-y_1^4)^2 + x_1^2y_1(x_1+y_1^2)^2 = 0\}$ is still singular and intersects D_1 at the point $x_1 = y_1 = 0$. Since this point only occurs in the (x_1,y_1) chart, we discard the other chart. We draw this schematically (drawing a single curve for V_1 is a convenience which is not meant to prejudice its reducibility or irreducibility):

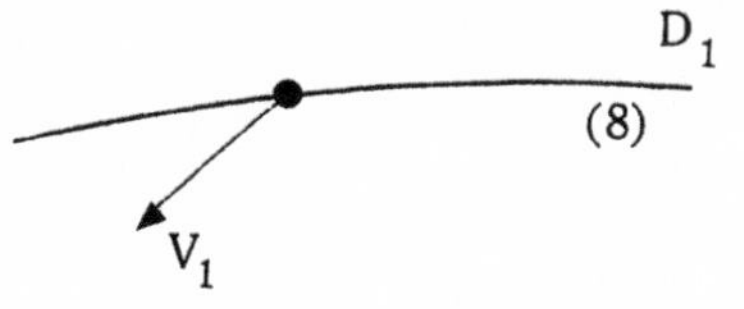

We now blow up at $x_1 = y_1 = 0$, considering only the coordinate system (x_2,y_2) for which $\pi_2 : \widetilde{U}_2 \to \widetilde{U}_1$ is given by $\pi_2(x_2,y_2) = (x_1,y_1) = (x_2y_2,y_2)$, since nothing of interest happens in the other charts. Then $\widetilde{f}_2 = \widetilde{f}_1 \circ \pi_2$ is given by

$$\begin{aligned}\widetilde{f}_2(x_2,y_2) &= (x_2y_2)^8[((x_2y_2)^2-y_2^4)^2 + (x_2y_2)^2y_2(x_2y_2+y_2^2)^2] \\ &= x_2^8y_2^{12}[(x_2^2-y_2^2)^2 + x_2^2y_2(x_2+y_2)^2] .\end{aligned}$$

$D_2 = \{y_2=0\}$ occurs with multiplicity 12 in $\widetilde{f}_2^{-1}(0)$. Schematically $\widetilde{f}_2^{-1}(0)$ looks like:

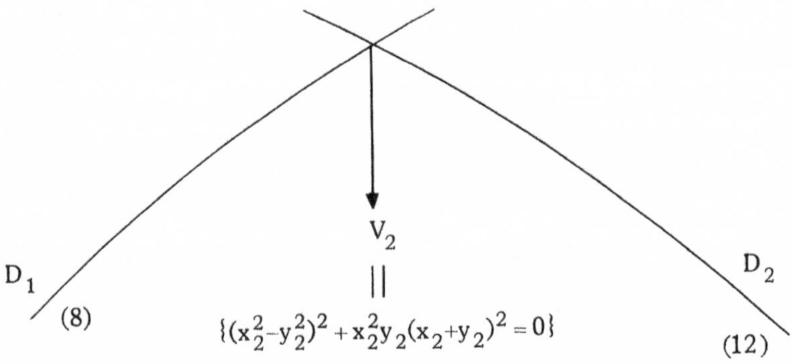

Blow up again; either chart can be used, so we use $\pi_3 : \widetilde{U}_3 \to \widetilde{U}_2$ given by $(x_2,y_2) = (x_3,x_3y_3)$.

$$\widetilde{f}_3(x_3,y_3) = x_3^{24}y_3^{12}[((1-y_3)^2 + x_3y_3)(1+y_3)^2] .$$

The proper transform $V_3 = \{((1-y_3)^2 + x_3y_3)(1+y_3)^2 = 0\}$ meets $D_3 = \{x_3=0\}$ at $y_3 = \pm 1$. It consists of two disjoint components. At $y_3 = -1$, V_3 is given by $(1+y_3)^2 = 0$, up to a nonzero factor, so this component, V_3^-, is nonsingular with multiplicity 2, meeting D_3 transversely. At $y_3 = +1$, V_3 is nonsingular but still tangent to D_3. Call this component V_3^+.

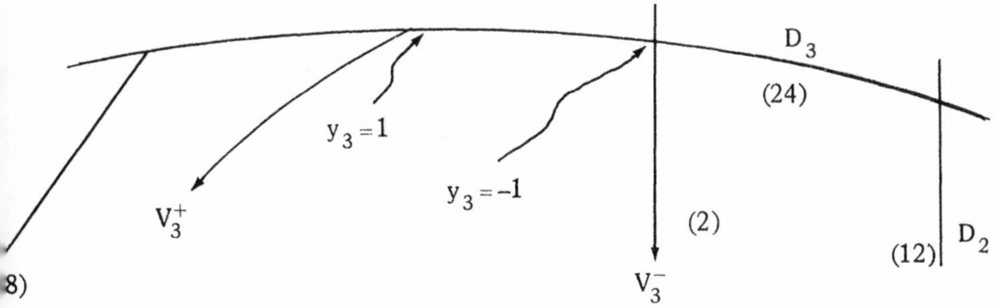

Changing coordinates by $\bar{x}_3 = x_3$, $\bar{y}_3 = y_3 - 1$, to center at the point (0,1), $\tilde{f}_3$ is given (up to the nonzero factor $y_3^{12}(1+y_3)^2$) by

$$\tilde{f}_3(\bar{x}_3,\bar{y}_3) = \bar{x}_3^{24}[\bar{y}_3^2+\bar{x}_3\bar{y}_3+\bar{x}_3]$$

Blow up at $\bar{x}_3 = \bar{y}_3 = 0$ using $(\bar{x}_3,\bar{y}_3) = (x_4y_4,y_4)$:

$$\tilde{f}_4(x_4,y_4) = x_4^{24}y_4^{25}[y_4+x_4y_4+x_4]$$

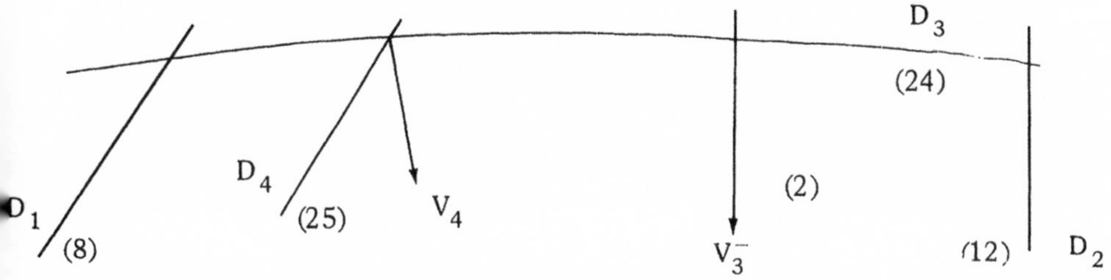

We still have a triple intersection point at $x_4 = y_4 = 0$. Blowing up once more completes the resolution: $(x_4,y_4) = (x_5y_5,y_5)$,

$$\tilde{f}_5(x_5,y_5) = x_5^{24}y_5^{50}(1+x_5y_5+x_5) .$$

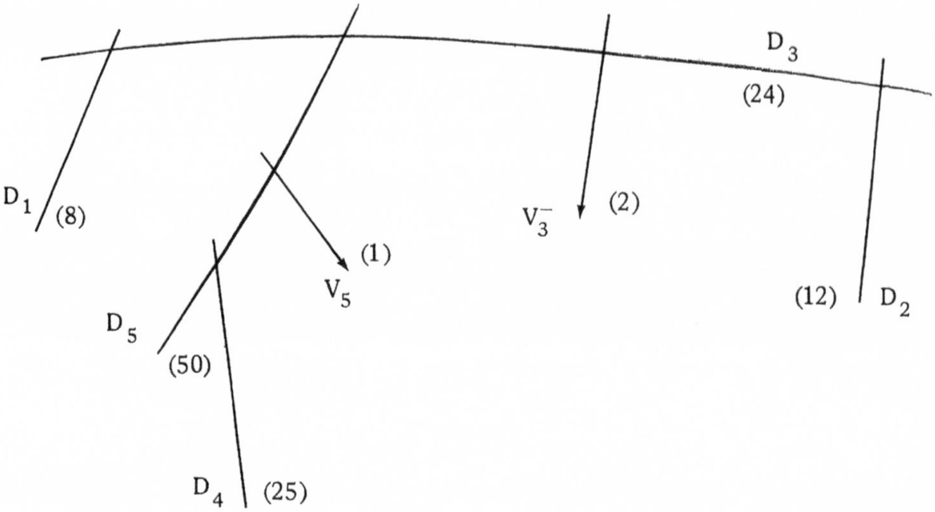

Here $V_5 = \{1+x_5y_5+x_5=0\}$ is nonsingular of multiplicity 1 and meets $D_5 = \{y_5=0\}$ at $x_5 = -1$ transversely. Thus this is the final resolution picture. We represent it by the dual "plumbing diagram"

$\Lambda =$

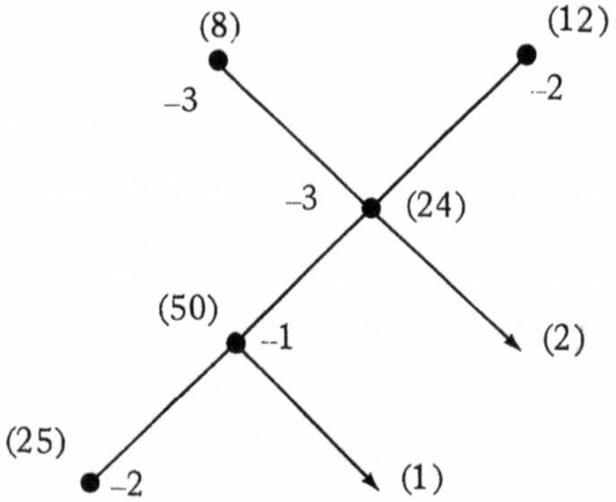

in which we have also written in the self-intersection numbers $e_i = D_i \cdot D_i$ of the exceptional curves. The e_i can be computed during the resolution process: when an exceptional curve first appears it appears with $D_i \cdot D_i = -1$ and each time a point is blown up on it $D_i \cdot D_i$ decreases by 1. Only the e_i

and not the multiplicities ℓ_i of the D_i, are relevant in considering Δ as a plumbing graph.

Thus $L(f)$ can be represented by plumbing according to the above graph. As we shall soon see (Theorem 20.1), it is easy to deduce the splice diagram for $L(f)$. The answer in this case is

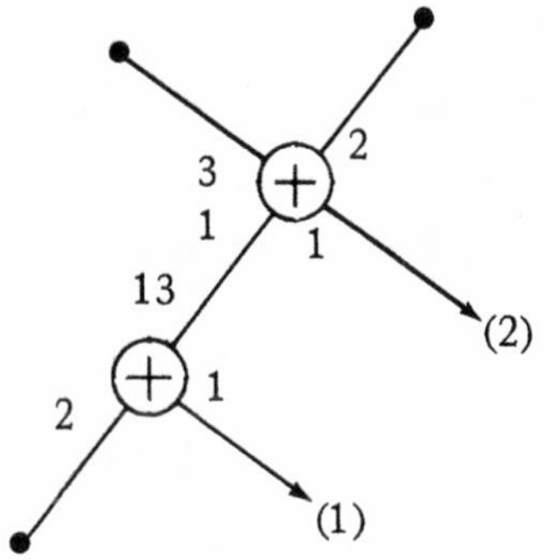

It is also easy to recover the multiplicities ℓ_i from the e_i and the "link component multiplicities" (1 and 2 in the above example). Namely let $A = (D_i \cdot D_j)$. That is

$$A_{ij} = \begin{cases} e_i & i = j \\ 1 & i \text{ connected to } j \text{ by an edge of } \Delta \\ 0 & \text{otherwise} . \end{cases}$$

Then A is invertible and

$$A \begin{pmatrix} \ell_1 \\ \vdots \\ \ell_n \end{pmatrix} + \begin{pmatrix} s_1 \\ \vdots \\ s_n \end{pmatrix} = 0$$

where s_i is the sum of the multiplicities of those components of the proper transform $\tilde{V}$ which intersect D_i. For instance

$$\begin{pmatrix} -3 & 0 & 1 & 0 & 0 \\ 0 & -2 & 1 & 0 & 0 \\ 1 & 1 & -3 & 0 & 1 \\ 0 & 0 & 0 & -2 & 1 \\ 0 & 0 & 1 & 1 & -1 \end{pmatrix} \begin{pmatrix} 8 \\ 12 \\ 24 \\ 25 \\ 50 \end{pmatrix} + \begin{pmatrix} 0 \\ 0 \\ 2 \\ 0 \\ 1 \end{pmatrix} = 0$$

in the above example.

This is, of course, just Theorem 18.2, but an alternate proof in the present context can be given by observing that, in our earlier notation, $f^{-1}(0)$ (with multiplicities) is homologically the same as $f^{-1}(\delta)$ for $0 \neq \delta$ small, and hence has zero intersection number with each D_i.

20. *From plumbing to splice diagram*

Let Δ be a plumbing diagram for a graph link, possibly with multiplicities $\underline{m}$ also given, so Δ represents a multilink.

THEOREM 20.1. *A splice diagram* Γ *for this link or multilink can be constructed as described below. If* Δ *is a minimal plumbing diagram then the resulting* Γ *is a minimal splice diagram.*

A chain in Δ is a portion of Δ of the form

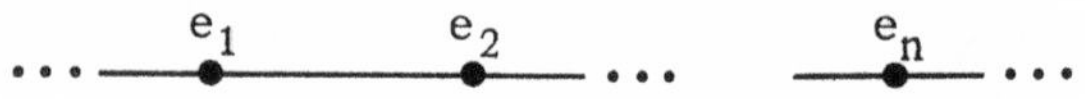

$(n \geq 0)$.

STEP 1. Make $\det(-A(\Delta)) = +1$. If this is not already so, it can be done by a single (+1)-blow-up on some chain of Δ.

STEP 2. The underlying graph of Γ is obtained from Δ by replacing each maximal chain by a single edge. All nodes are weighted +1 .

STEP 3. The weight at the end of an edge of Γ is $\det(-A(\Delta_1))$, where Δ_1 is the subgraph of Δ cut off by the corresponding edge of Δ (see figure). If Δ_1 is a single arrowhead we define $\det(-A(\Delta_1)) = 1$.

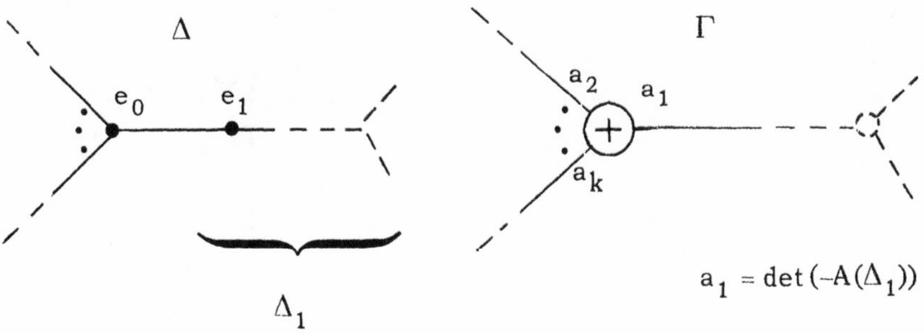

STEP 4. Multiplicities, if relevant, equal the corresponding multiplicities in Δ.

REMARK 1. Step 1 can be dispensed with, but then, if $\det(-A(\Delta)) = -1$, Steps 2 to 4 yield a splice diagram for the given link or multilink with ambient orientation reversed.

REMARK 2. For any plumbing graph Δ, if $-\Delta$ is the same graph with weights reversed in sign, then

$$\det(-A(\Delta)) = \det(A(-\Delta)) .$$

Indeed $-A(\Delta) = DA(-\Delta)D$ for a suitable diagonal matrix D with diagonal entries ± 1.

An algorithm for computing such determinants is given in the next section.

Examples. The reader is invited to verify the following examples.

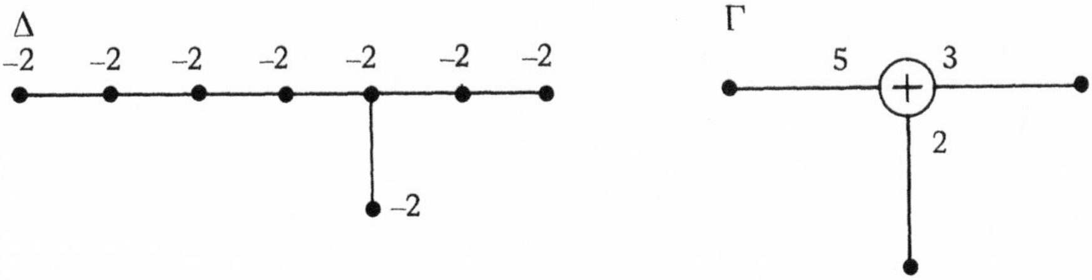

Δ

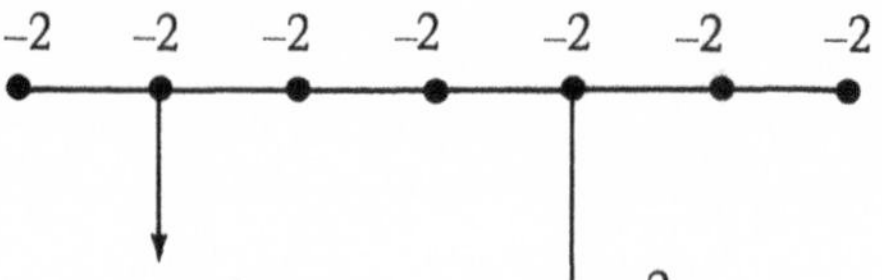

Γ

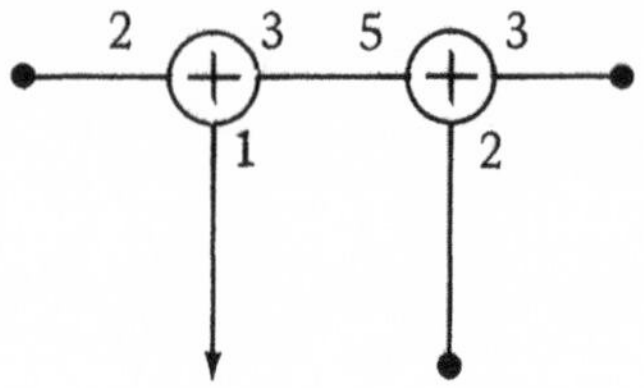

Δ

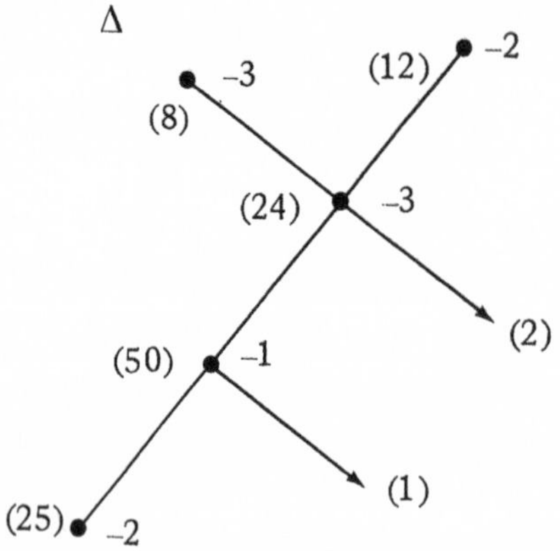

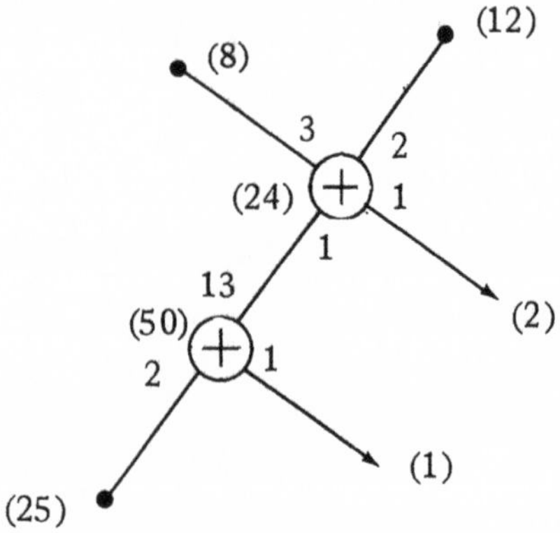

Proof of Theorem 20.1. The statement implicit in "Step 2" and the second sentence of the theorem follows from section 5 of [Ne 3] which describes how plumbing relates to the Jaco-Shalen-Johannsen splitting of a graph manifold (or equivalently to Waldhausen's classification of such manifolds). Thus we must only verify the claim about the numerical weights for Γ.

If Δ is a plumbing tree, let us denote, for the purpose of this proof, $B(\Delta) = -A(\Delta)$ and $d(\Delta) = \det(B(\Delta))$.

LEMMA 20.2. *The adjoint* $\mathrm{Adj}(B(\Delta)) = (B_{ij})$ *is given by*

$$B_{ji} = B_{ij} = \prod_{k=1}^{s} d(\Delta_k)$$

where $\Delta_1, \cdots, \Delta_s$ *are the components of the graph obtained from* Δ *by deleting the path in* Δ *from* i *to* j *and all incident edges (of course,* s *and* $\Delta_1, \cdots, \Delta_s$ *depend on* (i,j)).

Proof. By renumbering the vertices of Δ we can assume $i = 1$, that the path from i to j is $(1,2,\cdots,j)$, and that $B(\Delta)$ has the following form.

$$B(\Delta) = \left(\begin{array}{ccccc|cccc}
b_1 & -1 & & & 0 & & & & \\
-1 & b_2 & -1 & & & & & C & \\
 & & \ddots & & -1 & & & & \\
0 & & & -1 & b_j & & & & \\ \hline
 & & & & & B(\Delta_1) & 0 & & \\
 & & & & & 0 & B(\Delta_2) & & 0 \\
 & & C^t & & & & & \ddots & 0 \\
 & & & & & & 0 & 0 & B(\Delta_s)
\end{array}\right)$$

Here C is zero except for s entries -1. The k-th such entry occurs above $B(\Delta_k)$ and in row n_k say. We may assume $1 \leq n_1 \leq n_2 \leq \cdots \leq n_s \leq j$.

The (1,j)-minor of $B(\Delta)$ is thus

$$B(\Delta)^{1,j} = \left(\begin{array}{ccccc|ccc}
-1 & b_2 & -1 & \cdots & & & & \\
0 & -1 & b_3 & \cdots & & & C' & \\
 & & \ddots & \ddots & b_{j-1} & & & \\
0 & & & & -1 & & & \\ \hline
 & & & & & B(\Delta_1) & & 0 \\
 & & * & & & & \ddots & \\
 & & & & & 0 & & B(\Delta_s)
\end{array}\right).$$

We do column operations to this: add multiples of the $(j-1)^{st}$ column to later columns to clear the $(j-1)^{st}$ row of C', then add multiples of the $(j-2)^{nd}$ column to later columns to clear the $(j-2)^{nd}$ row of C', and so on. Using the fact that $n_1 \leq n_2 \leq \cdots \leq n_s$ one sees that the result is of the form

$$\left(\begin{array}{c|c}
\begin{matrix} -1 & & * \\ & \ddots & \\ 0 & & -1 \end{matrix} & 0 \\ \hline
* & \begin{array}{c|c|c|c} B(\Delta_1) & * & & * \\ \hline 0 & B(\Delta_2) & & \\ & 0 & \ddots & \\ \hline & & 0 & B(\Delta_s) \end{array}
\end{array}\right)$$

which evidently has determinant $(-1)^{j-1} \prod_{k=1}^{s} d(\Delta_k)$. We must multiply this by $(-1)^{i+j} = (-1)^{1+j} = (-1)^{1-j}$ to get the corresponding entry of $\mathrm{Adj}(B(\Delta))$, so the lemma is proved.

Returning to the proof of Theorem 20.1, we assume "Step 1" done, so $\det(B(\Delta)) = 1$. Then the formula of Theorem 18.2 becomes $\vec{\ell} = \mathrm{Adj}(B(\Delta))\vec{s}$.

This implies that the linking number $\ell(S_i, S_j)$ of a fiber S_i at the i-th vertex of Δ with a fiber S_j at the j-th vertex of Δ is $B_{ij} = \prod_k d(\Delta_k)$, as given in Lemma 20.2. (We may have to add an arrow to Δ at vertex i to represent S_i in order to apply Theorem 18.2.)

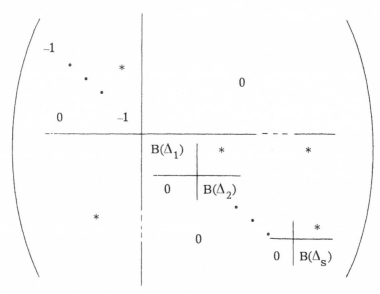

This computation of $\ell(S_i, S_j)$ agrees with Theorem 10.1, if we assign weights to Γ as claimed in our present theorem. But it is an easy induction to see that the knowledge of $\ell(S_i,S_j)$ for all i,j determines the weights of Γ, so Theorem 20.1 follows. We must include the case $i=j$ here, in which case we mean the linking number $\ell(S_i,S_i')$ of two different fibers at vertex i (by adding an arrow to Γ at vertex i, Theorem 10.1 computes this as the product of all the edge weights adjacent to this vertex).

21. *The determinant of a graph*

This section is mostly a paraphrase of part of the Dissertation of N. Duchon (Maryland, 1982).

We describe a simple algorithm to diagonalize matrices of the form $A(\Delta)$, where Δ is a plumbing tree. The diagonalized matrix has the form $D = P^tA(\Delta)P$ with $\det(P) = 1$. Thus $\det(D) = \det(A(\Delta))$ and $\operatorname{sign}(D) = \operatorname{sign}(A(\Delta))$.

We must allow trees weighted by arbitrary rational numbers e_i rather than just by integers. Given such a tree, pick some vertex and direct all edges toward this vertex. Now simplify Δ by recursively deleting edges according to the following procedures (the portion of Δ drawn to the left of e_j may be absent; we have also reindexed as necessary):

$(e_i \neq 0,\ i=1,\cdots,k)$

$$e_j' = e_j - \frac{1}{e_1} - \cdots - \frac{1}{e_k}$$

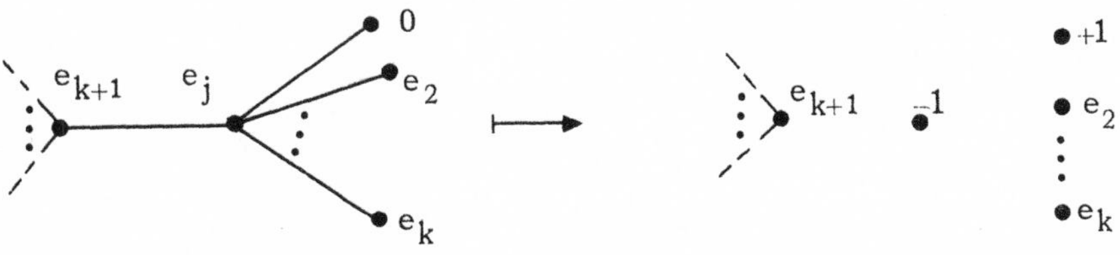

Eventually we end up with a collection of isolated points with weights $d_1, \cdots, d_n$ say. Then $D = \mathrm{diag}(d_1, \cdots, d_n)$ is the desired diagonalization of $A(\Delta)$.

Example.

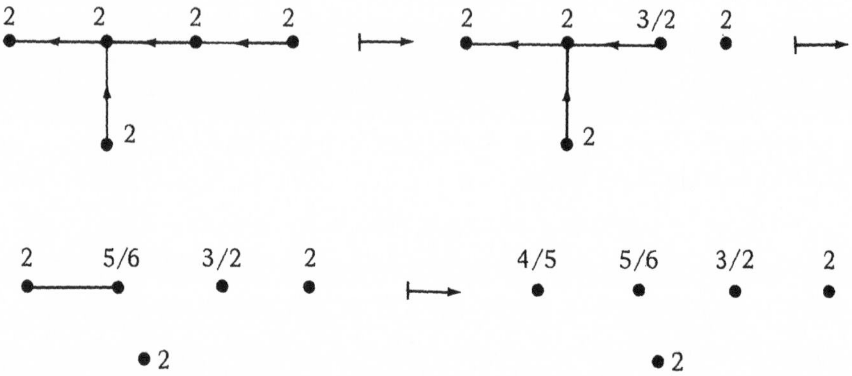

Thus $\det(A(\Delta)) = \frac{4}{5} \cdot \frac{5}{6} \cdot \frac{3}{2} \cdot 2 \cdot 2 = 4$.

The proof of this algorithm consists of the observation that our procedures simply correspond to certain simultaneous row and column operations on $A(\Delta)$; we leave this to the reader.

In the above procedure let Δ_j be the portion of Δ to the right of, and including, the vertex weighted e_j, and define a number $\mathrm{cf}(\Delta_j, j) \in \mathbf{Q} \cup \{-1/0\}$ by

$$\mathrm{cf}(\Delta_j, j) = \begin{cases} e_j' \,, & \\ & \text{or} \\ -1/0 & \end{cases}$$

according as the step of the procedure which deletes the edges coming into j is of the first or second type.

Thus if $\mathrm{cf}(\Delta_j, j) \neq -1/0$ for all vertices of Δ, then the numbers $\mathrm{cf}(\Delta_j, j)$ are just the entries of our final diagonal matrix D, so

$$\det(A(\Delta)) = \prod_j cf(\Delta_j, j) .$$

this formula is clearly valid in general if we make the convention: $0 \cdot (-1/0) = -1$.

Since $\Delta_j = \Delta$ if and only if j is the vertex toward which we have directed Δ, the notation of $cf(\Delta, j)$ implicitly assumes Δ has been directed toward j. The notation

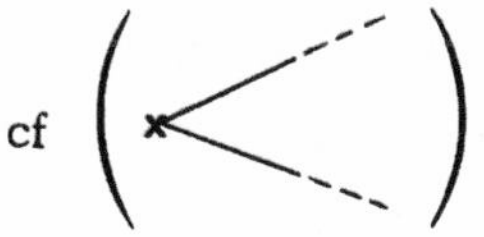

means the vertex marked with a cross is the chosen one.

Examples.

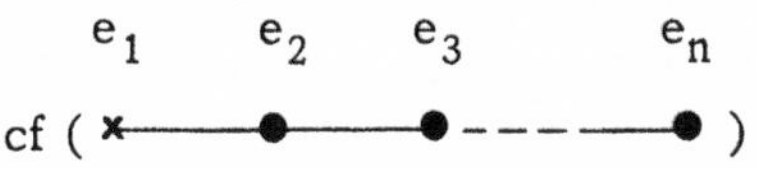

$$e_1 - \cfrac{1}{e_2 - \cfrac{1}{\ddots - \cfrac{1}{e_n}}}$$

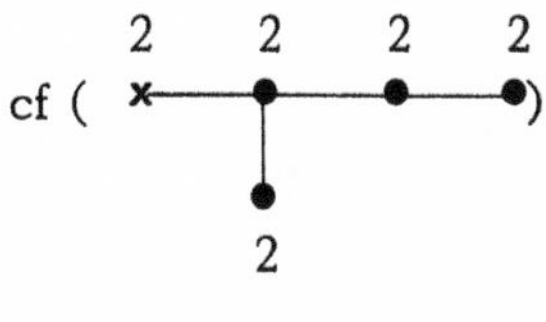

$$= \frac{4}{5}$$

$cf(\Delta, j)$ is a generalization of continued fractions, hence the notation "cf."

We shall need the following fact later.

PROPOSITION 21.1. *Let* Δ *be the union of two integrally weighted trees* Δ_0 *and* Δ_1 *intersecting in a single vertex* j :

$\Delta =$ 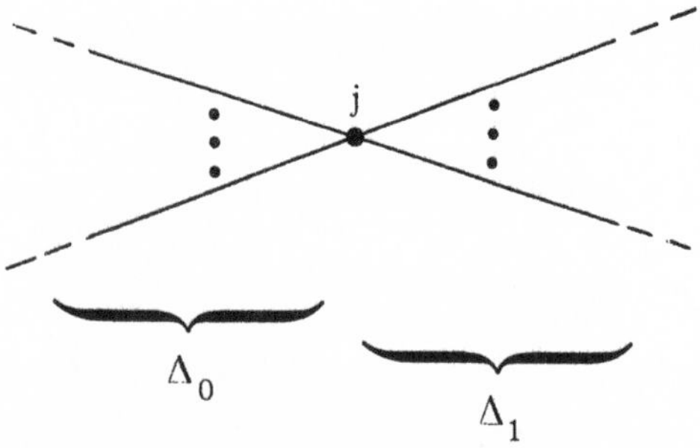

Suppose $cf(\Delta_0, j) = p/q$ *in lowest terms. Then*

(i) $p | \det A(\Delta_0)$, ($|$ denotes "divides"),

(ii) $(\det A(\Delta_0)/p) | \det A(\Delta)$.

In particular if $\det A(\Delta) = \pm 1$ *then* $\det A(\Delta_0) = \pm p$.

We omit the proof, which is an easy induction. Note that if Δ is the resolution graph of a plane curve singularity then $-A(\Delta)$ is positive definite, so with notation as in the proposition, $\det(-A(\Delta_0)) = |p|$. This computes the weights in the corresponding splice diagram completely in terms of generalized continued fractions (see Theorem 20.1).

22. *From splice diagrams to plumbing*

In order to construct a plumbing representation for a graph link given by a splice diagram we need two things:

(i) Plumbing diagrams for the basic building blocks: Seifert links;

(ii) A procedure to splice together plumbing diagrams.

The first is given by a minor modification of a theorem of von Randow [vR], namely "Algorithm 5.4" of [N-R]:

THEOREM 22.1. *The Seifert link with splice diagram*

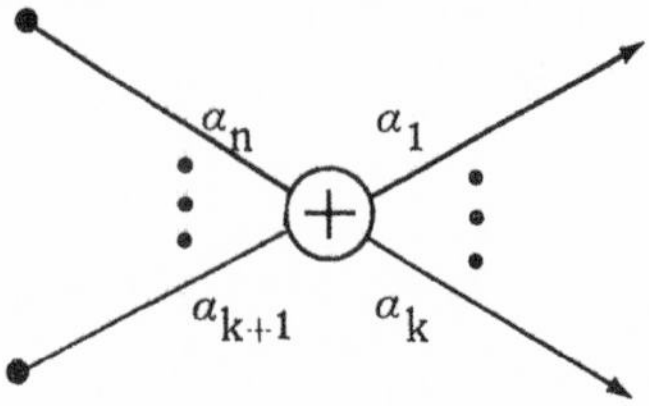

has normal form plumbing diagram

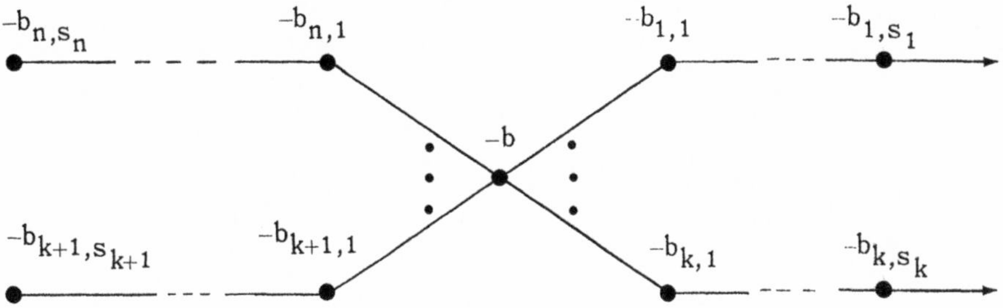

with weights determined by the following conditions:

$$b_{ij} \geq 2 \quad \textit{for all} \quad i,j$$

$$\frac{\alpha_1 \cdots \hat{\alpha}_i \cdots \alpha_n}{\alpha_i} = \mathrm{cf}\left(\overset{b_{i,1}}{\bullet}\!\!-\!\!\overset{b_{i,2}}{\bullet}\!\!-\!-\!-\!\!\overset{b_{i,s_i}}{\bullet}\!\!-\!\!\overset{b_i}{\times}\right) \quad \textit{with} \quad b_i \geq 1$$

$$b = \sum_{i=1}^{n} \frac{1}{\mathrm{cf}\left(\overset{b_{i,1}}{\times}\!\!-\!-\!-\!\!\overset{b_{i,s_i}}{\bullet}\right)} + \frac{1}{\alpha_1 \cdots \alpha_n} .$$

The intersection form for this plumbing diagram is negative definite.

Note that $b_i = \left\lceil \frac{\alpha_1 \cdots \hat{\alpha}_i \cdots \alpha_n}{\alpha_i} \right\rceil$, where $\lceil\ \ \rceil$ is "least integer greater than or equal to." These numbers have a significance that will be important in Theorem 22.2. To describe it, let Δ' be the above plumbing diagram with arrowheads deleted and v the outermost vertex of the i-th branch of Δ' (the vertex weighted $-b_{i,s_i}$).

ADDENDUM TO 22.1. $\mathrm{cf}(\Delta,v) = \frac{1}{-b_i}$.

Proof. Let Δ'' be obtained from Δ' by attaching an extra vertex w with weight $-b_i$ at the end of the i-th branch:

$\Delta'' =$

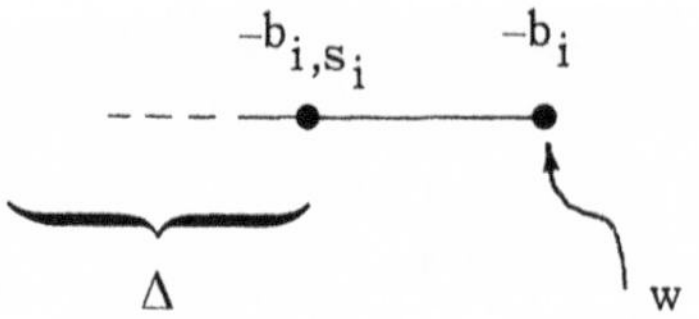

Now let v_0 be the central node of Δ''. Then

$$\mathrm{cf}(\Delta',v) = \frac{1}{-b_i} \Longleftrightarrow \mathrm{cf}(\Delta'',w) = 0 \Longleftrightarrow \det A(\Delta'') = 0 \Longleftrightarrow \mathrm{cf}(\Delta'',v_0) = 0\ ,$$

where the last two equivalences are by Proposition 21. The equation $\mathrm{cf}(\Delta'',v_0) = 0$ can be verified by a simple computation using the fact that,

$$\text{if } \mathrm{cf}\left(\overset{b_1}{\times}\!\!-\!\!\overset{b_2}{\bullet}\!\!-\,-\,-\!\!\overset{b_n}{\bullet}\right) = \frac{p}{q}\,, \quad \text{then } \mathrm{cf}\left(\overset{b_1}{\bullet}\!\!-\!\!\overset{b_2}{\bullet}\!\!-\,-\,-\!\!\overset{b_n}{\times}\right) = \frac{p}{q'} \quad \text{with}$$

$qq' \equiv 1 \bmod p$.

We next address the problem of splicing together plumbing diagrams. Let $L_1 = (\Sigma_1,K_1)$ and $L_2 = (\Sigma_2,K_2)$ be graph links given by plumbing according to

We shall splice along the link components corresponding to the indicated arrowheads.

THEOREM 22.2. *Let* $\frac{1}{a_1} = \mathrm{cf}(\Delta_1',v)$, $\frac{1}{a_2} = \mathrm{cf}(\Delta_2',w)$.

Then the following is a plumbing diagram for the spliced link:

$\Delta =$

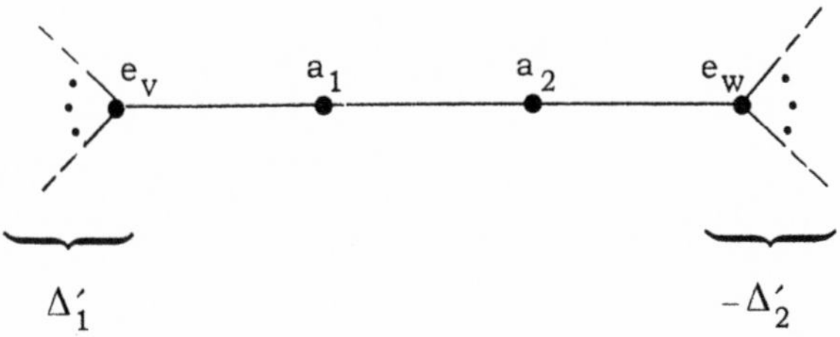

where the $-\Delta_2'$ *indicates that multiplicities, if given, should be reversed in sign in this half of the diagram.*

Proof. In view of part 3 of Theorem 18.3, Δ is equivalent to the diagram

$\bar{\Delta} =$

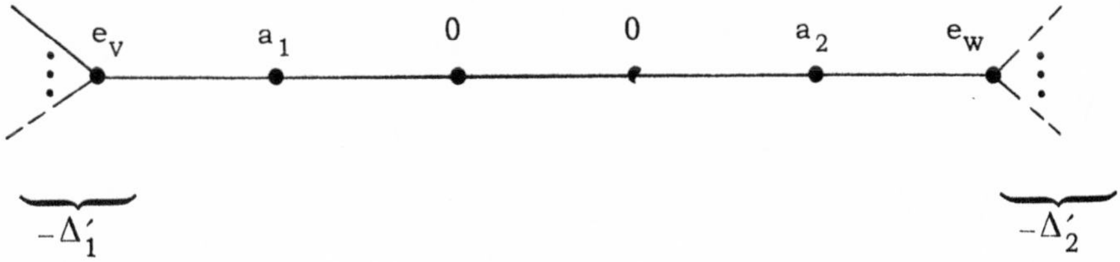

Moreover the graphs

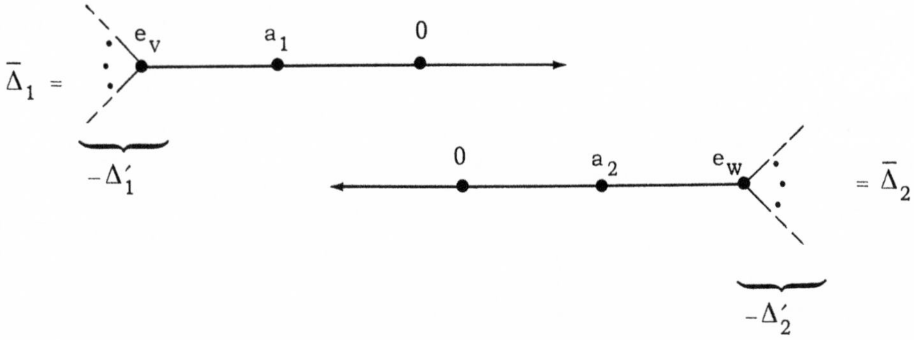

are equivalent to Δ_1 and Δ_2 respectively.

Now

cf 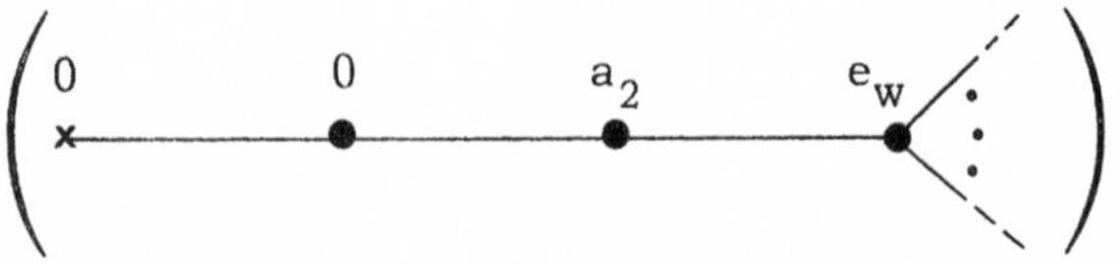 = 0

whence follows that for any vertex j of $\bar{\Delta}_1$,

$$\mathrm{cf}(\bar{\Delta}_1, j) = \mathrm{cf}(\bar{\Delta}, j) .$$

Similarly

$$\mathrm{cf}(\bar{\Delta}_2, j) = \mathrm{cf}(\bar{\Delta}, j) .$$

for any j in $\bar{\Delta}_2$. Theorem 22.2 thus follows from Theorem 20.1 and equation (21.1).

Example. By Theorem 22.1, the following two splice diagrams

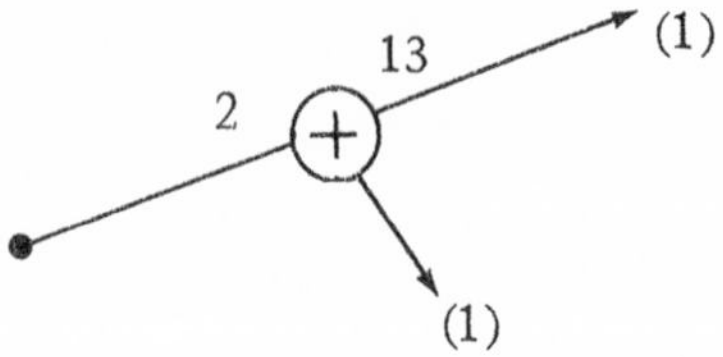

and

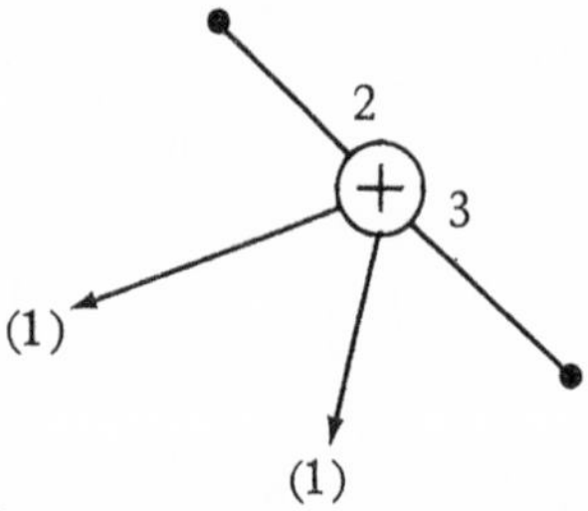

(omitted weights are 1)

have normal form plumbing diagrams

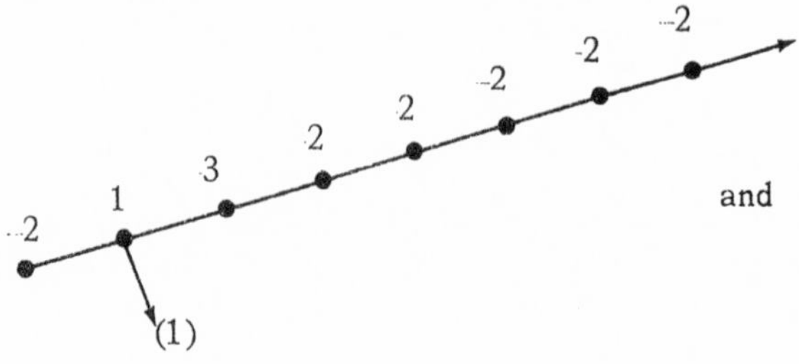

and

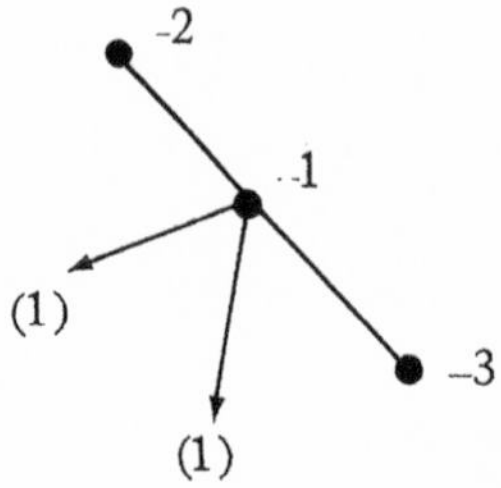

Moreover, by the Addendum to 22.1:

$$\mathrm{cf}\left(\overset{-2}{\bullet}\!-\!\!-\!\overset{-1}{\times}\!-\!\!-\!\overset{-3}{\bullet}\right) = \frac{1}{-b_1} \quad \text{with} \quad b_1 = \left\lceil \frac{2\cdot 3}{1} \right\rceil = 6\ ,$$

$$\mathrm{cf}\left(\overset{-2}{\bullet}\!-\!\!-\!\overset{-1}{\bullet}\!-\!\!-\!\overset{-3}{\bullet}\!-\!\!-\!\overset{-2}{\bullet}\!-\!\!-\!\overset{-2}{\bullet}\!-\!\!-\!\overset{-2}{\bullet}\!-\!\!-\!\overset{-2}{\bullet}\!-\!\!-\!\overset{-2}{\times}\right) = \frac{1}{-b_2} \quad \text{with} \quad b_2 = \left\lceil \frac{2}{13} \right\rceil = 1\ .$$

Thus, by Theorem 22.2,

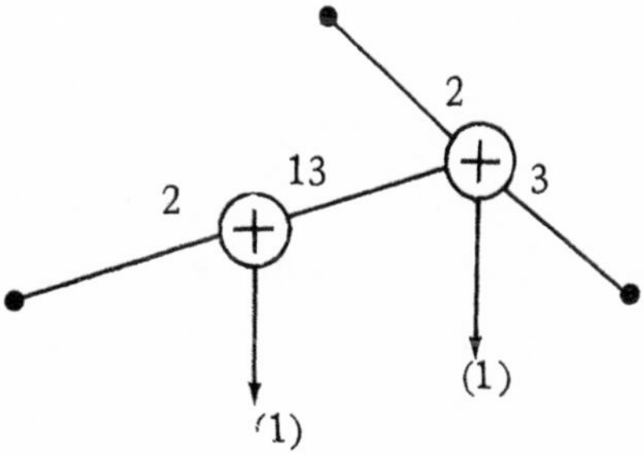

has plumbing diagram

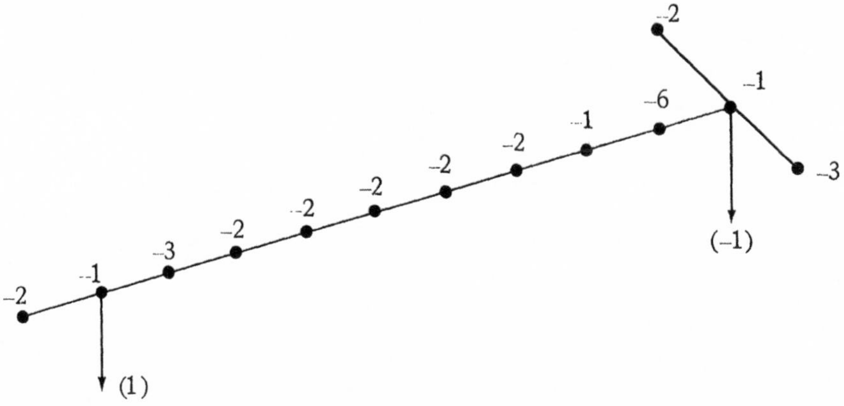

By six blow-downs and a zero-chain absorption this simplifies to

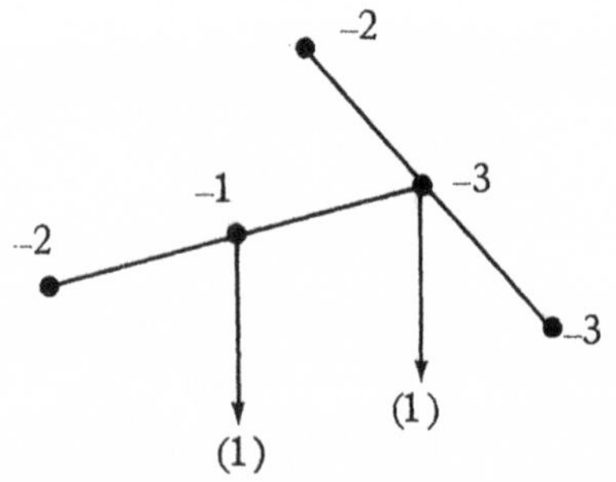

23. *Invariants via plumbing*

We can think of a plumbing diagram for a graph multilink as a (far from minimal) splice diagram: the vertices of the plumbing diagram correspond to Seifert pieces in a decomposition of the link exterior, in fact to pieces of the form $X \times S^1$. The multiplicity at a vertex can be interpreted as the linking number of an S^1 fiber in $X \times S^1$ with the multilink. The results of sections 11 and 12 can thus be applied directly to give the following theorems (for 23.1 see also [Ev]; for 23.3 see also [A'C]).

Let the arrow-head vertices of Δ be $v_1, \cdots, v_n$. By Theorem 18.2, there exist integers ℓ_{vi}, $v \in \overline{\mathfrak{A}}(\Delta)$, such that, if vertex v_i is assigned multiplicity m_i for each i then the multiplicity at $v \in \overline{\mathfrak{A}}(\Delta)$ is

$$\ell_v = \sum_{i=1}^{n} \ell_{vi} m_i .$$

THEOREM 23.1. $\Delta^L(t_1, \cdots, t_n) = \prod_{v \in \overline{\mathfrak{A}}} (t_1^{\ell_{v1}} t_2^{\ell_{v2}} \cdots t_n^{\ell_{vn}} - 1)^{\delta_v - 2}, \quad n \geq 2$

$$= (t_1 - 1) \prod_{v \in \overline{\mathfrak{A}}} (t_1^{\ell_{v1}} - 1)^{\delta_v - 2} , \quad n = 1 .$$

Here terms of the form $(t_1^0 \cdots t_n^0 - 1)^0$ *should be formally cancelled before being set equal to zero.*

THEOREM 23.2. *$L(\underline{m})$ is fibered if the multiplicity of each node is non-zero. If the plumbing graph is minimal (see section 20) this condition for fiberability is both necessary and sufficient.*

This theorem remains valid even if the ambient space is not a homology sphere.

Let $d = \gcd(m_1, \cdots, m_n)$.

THEOREM 23.3. *If $L(\underline{m})$ is fibered then the characteristic polynomial of the monodromy is*

$$\Delta_1(t) = (t^d - 1) \prod_{v \in \overline{\mathfrak{A}}} (t^{\ell_v} - 1)^{\delta_v - 2} .$$

The results of section 14 can also be modified for plumbing diagrams. Let $\mathfrak{E}$ be the set of edges of Δ which join two non-arrowhead vertices of Δ. For $v \in \overline{\mathfrak{A}}$ denote by d_v the g.c.d. of all multiplicities at vertices of Δ equal to or adjacent to v. For $E \in \mathfrak{E}$ denote by d_E the g.c.d. of the multiplicities at the endpoints of E.

THEOREM 23.4. *Suppose the multiplicities ℓ_v of all $v \in \overline{\mathfrak{A}}(\Delta)$ are positive (so, in particular, $L(\Delta(\underline{m}))$ is fibered), and let q be a common multiple of the orders of the eigenvalues of the monodromy h_*. Then the characteristic polynomial $\Delta'(t)$ of $h_* | (h_*^q - 1) H_1(F)$ is*

$$\Delta'(t) = (t^d - 1) \prod_{E \in \mathfrak{E}} (t^{d_E} - 1) / \prod_{v \in \overline{\mathfrak{A}}} (t^{d_v} - 1) .$$

Moreover, if $\mathfrak{N}'$ is the set of "separating nodes" of Δ, that is, nodes v of Δ with arrowheads in at least two components of $\Delta - \{v\}$, and if $\mathfrak{E}'$ is a set of edges consisting of just one edge from each chain connecting two separating nodes, then

$$\Delta'(t) = \begin{cases} (t^d-1) \prod_{E \in \mathcal{E}'} (t^{d_E}-1) / \prod_{v \in \mathcal{N}'} (t^{d_v}-1), & \mathcal{N}' \neq \emptyset , \\ 1 & , \mathcal{N}' = \emptyset . \end{cases}$$

The assumption of positive multiplicities in this theorem replaces the uniform twists assumption of Theorem 14.1. Indeed, it is not hard to verify (see [Ne 4]) that if E is an edge of Δ joining vertices v and w with multiplicities ℓ_v and ℓ_w, then there are $d_E = \gcd(\ell_v, \ell_w)$ annuli in the fiber F corresponding to this edge and the twist on any one of them is $-d_E/\ell_v\ell_w$ (see section 13 for terminology). The twist for a single edge of the corresponding splice diagram is obtained by summing the $-d_E/\ell_v\ell_w$ over all edges E of a maximal chain in Δ.

In the resolution graph of a plane curve singularity all the multiplicities (even the m_i) are positive, so Theorem 23.4 applies.

In [Ne 4] it is described how to go further and compute the "Jordan normal form" of the real Seifert form of an algebraic link from its resolution diagram and the same is done in terms of the splice diagram in [Ne 5].

24. *Criteria for algebraicity*

Define an *algebraic multi-link* to be a graph multi-link which is the link of a pair (V,C) consisting of an algebraic curve C in a normal algebraic surface C at a point $p \in C$ which is possibly singular for both.

Let $\Delta(\underline{m})$ be a plumbing graph for a graph multi-link.

THEOREM 24.1. *The following statements are equivalent:*

(i) $\Delta(\underline{m})$ *is a resolution graph of an algebraic multi-link;*

(ii) *The intersection form* $A(\Delta)$ *is negative definite and the multiplicities* m_i, $i \in \mathcal{A}(\Delta)$ *are positive;*

(iii) *All euler weights* e_v, $v \in \overline{\mathcal{A}}(\Delta)$ *are negative and all multiplicities* ℓ_v, $v \in \overline{\mathcal{A}}(\Delta)$ *and* m_i, $i \in \mathcal{A}(\Delta)$ *are positive.*

Proof. That (ii) implies (i) is Grauert's criterion [Gr, p. 367], that (i) implies (iii) is clear, and that (iii) implies (ii) is Mumford's proof of DuVal's theorem ([Mu], see also [H-N-K, p. 86]).

In section 9 a criterion for algebraicity in terms of the splice diagram Γ was given; namely that all weights are positive and that in any portion

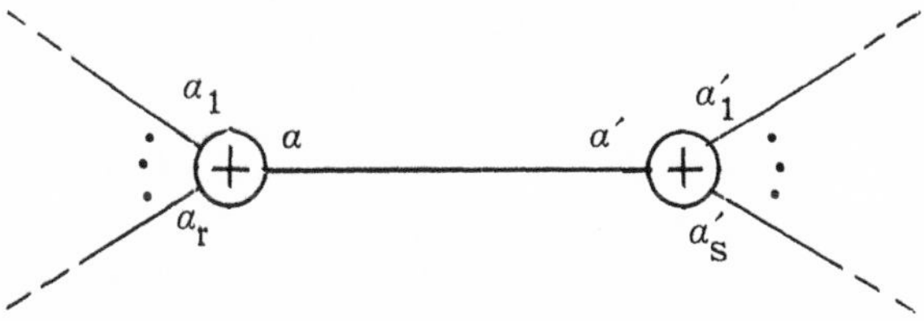

of Γ one has:

(*) $$a\,a' - \prod_1^r a_i \prod_1^s a'_j > 0\,.$$

We shall now describe a proof of this.

Necessity of the conditions. Construct the splice diagram Γ from a resolution diagram Δ as described in Theorem 20.1. Then Γ has positive weights by construction, since $A(\Delta)$ is negative definite, whence $\det(-A(\Delta_1)) > 0$ for any subdiagram of Δ. Moreover, condition (*) follows by comparing the computations of twists in Theorem 13.1 and section 23.

Sufficiency of the conditions. We treat first the case

Γ =

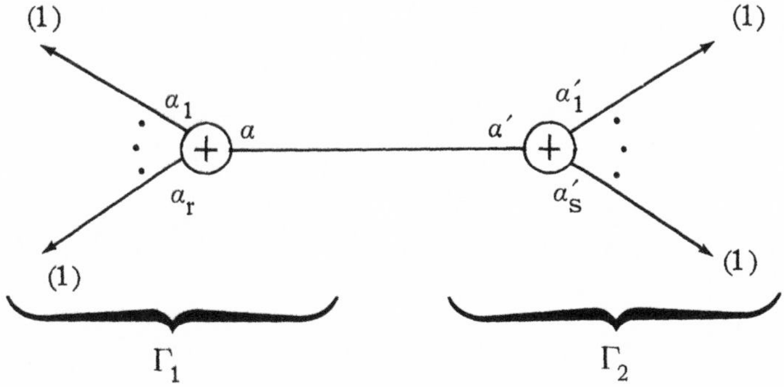

Construct a corresponding plumbing graph Δ by Theorems 22.1 and 22.2:

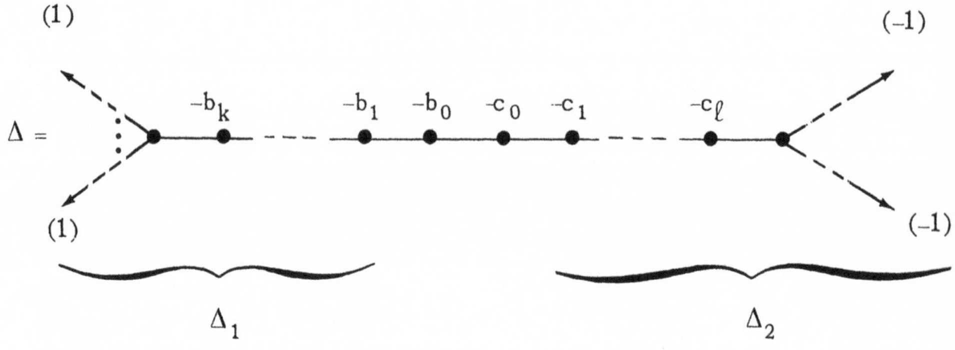

Thus Δ_1 and Δ_2 are the normal form plumbing graphs for Γ_1 and Γ_2 of Theorem 22.1 and b_0 and c_0 are as in the Addendum to Theorem 22.1. We claim:

(1) The intersection form $A(\Delta)$ has index $(1, n-1)$ where n is the number of non-arrowhead vertices of Δ (that is, $A(\Delta)$ diagonalizes over $\mathbf{R}$ to $\mathrm{diag}(1, -1, \cdots, -1)$.

(2) $\mathrm{cf}\begin{pmatrix} b_0 & b_1 & & b_k \\ \times & \bullet & \cdots & \bullet \end{pmatrix} \cdot \mathrm{cf}\begin{pmatrix} c_0 & c_1 & & c_\ell \\ \times & \bullet & \cdots & \bullet \end{pmatrix} < 1 .$

(3) $b_0 \geq 1,\ c_0 \geq 1,\ b_i \geq 2\ (i>1),\ c_i \geq 2\ (i>1)$.

Indeed, (3) is by construction. Property (1) follows from the fact that $A(\Delta_1)$ and $A(\Delta_2)$ are negative definite: if we diagonalize $A(\Delta)$ by the procedure of section 21 towards the $(-b_0)$-vertex, we end up with the entries of the diagonalizations of $A(\Delta_1)$ and $A(\Delta_2)$ and a "1" and a "-1." Property (2) is just a reformulation of equation (*), since

$$\mathrm{cf}\begin{pmatrix} b_0 & b_1 & & b_k \\ \times & \bullet & \cdots & \bullet \end{pmatrix} = \frac{\alpha_1 \cdots \alpha_r}{\alpha} =: \frac{\beta}{\alpha} \quad \text{(notation)},$$

$$\mathrm{cf}\begin{pmatrix} c_0 & c_1 & & c_\ell \\ \times & \bullet & \cdots & \bullet \end{pmatrix} = \frac{\alpha_1' \cdots \alpha_s'}{\alpha'} =: \frac{\beta'}{\alpha'} \quad \text{(notation)}.$$

Now Properties (2) and (3) imply that at least one of b_0 and c_0 equals 1; this follows from the inequalities

$$1 \le b_0 = \left\lceil \frac{\beta}{\alpha} \right\rceil, \quad 1 \le c_0 = \left\lceil \frac{\beta'}{\alpha'} \right\rceil, \quad \text{and} \quad \frac{\beta}{\alpha} \cdot \frac{\beta'}{\alpha'} < 1 .$$

Without loss of generality, $b_0 = 1$. Blow down that vertex (Theorem 18.3) to get

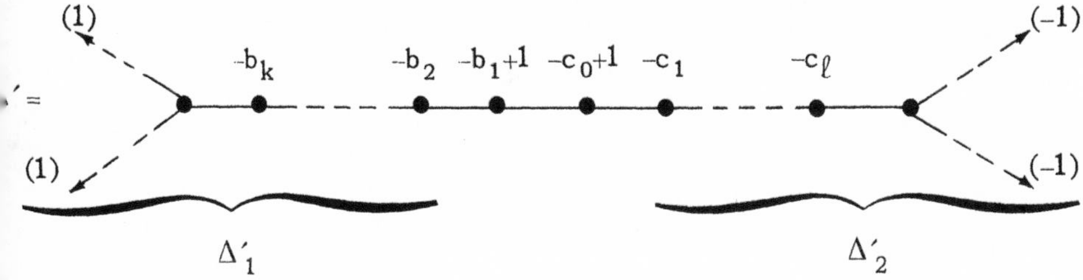

A (−1)-blow-down reduces the index of $A(\Delta)$ by $(0,1)$, so $A(\Delta')$ has index $(1, n-2)$. Thus Δ' still satisfies Property (1). It also still satisfies Property (2): elementary calculation shows

$$\mathrm{cf}\left(\overset{b_1-1}{\times} \!\!-\!\!-\!\! \overset{b_2}{\bullet} \cdots \overset{b_k}{\bullet} \right) = \frac{\beta}{\alpha-\beta} ,$$

$$\mathrm{cf}\left(\overset{c_0-1}{\times} \!\!-\!\!-\!\! \overset{c_1}{\bullet} \cdots \overset{c_\ell}{\bullet} \right) = \frac{\beta'-\alpha'}{\alpha'} ,$$

and the inequality $\frac{\beta}{\alpha-\beta} \cdot \frac{\beta'-\alpha'}{\alpha'} < 1$ is clearly equivalent to $\frac{\beta}{\alpha} \cdot \frac{\beta'}{\alpha'} < 1$. Thus if Δ' also satisfies Property (3) we can rename Δ' as Δ, reindex the vertices, and iterate by blowing down again, and so on. We must thus just consider the eventually occurring case that Property (3) is violated for Δ'. This can only happen because $c_0 - 1 = 0$. But then a zero-chain absorption can be performed to get a diagram Δ''. This zero-chain absorption reduces the index by $(1,1)$, so $A(\Delta'')$ is negative definite. It also flips the signs of the multiplicities in one half (say the right half) of the diagram, so the multiplicities of Δ'' are all positive again. Δ'' is thus the resolution graph of an algebraic link by part (ii) of Theorem 24.1.

The above proof dealt with the case that Γ has just two nodes. If Γ has more nodes, we can assume inductively that the two pieces being spliced have already been given negative definite plumbing diagrams. The above argument then goes through with no change to show the same for Γ, since the argument depends only on generalized continued fractions computed at points between the two nodes in question and these are unchanged by splicing or reduction moves elsewhere in the graph.

REFERENCES

[A'C] A'Campo, N.: Sur la monodromie des singularités isolées d'hypersurfaces complexes, Inv. Math. 20 (1973), 147-169.

[A] Ashley, C. W.: *The Ashley book of knots*, Doubleday, Garden City (1944).

[B-L] Blank, S., and Laudenbach, F.: Isotopie des formes fermées en dimension 3, Inv. Math. 54 (1979), 103-177.

[B-S] Bonahon, F., and Siebenmann, L.: New Geometric splittings of classical knots (algebraic knots), (to appear).

[B] Brauner, K: Zur Geometrie der Funktionen zweier Veränderlichen: II-IV. Abh. Math. Sem. Hamburg, 6 (1928), 1-54.

[B-K] Brieskorn, E., and Knörrer, H.: *Ebene algebraische Kurven*, Birkhäuser, Boston (1981).

[B-E] Buchsbaum, D. A., and Eisenbud, D.: Some structure theorems for finite free resolutions, Adv. in Math. 12 (1974), 84-139.

[Bu 1] Burau, W.: Kennzeichnung der Schlauchknoten. Abh. Math. Sem. Hamburg, 9 (1932), 125-133.

[Bu 2] Burau, W.: Kennzeichnung der Schlauchverkettungen. Abh. Math. Sem. Hamburg, 10 (1934), 285-297.

[Ce] Cerf, J.: 1-formes fermées non singulières sur les variétés compactes de dimension 3, Seminaire Bourbaki, June 1981, Springer Lect. Notes in Math. 901 (1981), 205-219.

[De] Deligne, P.: Résumé des premiers exposés de A. Grothendieck, Exposé I in Groupes de Monodromie en Géométrie Algébrique, SGA 7, I. Lecture Notes in Mathematics 288 (Springer-Verlag, 1972).

[Du] Durfee, A.: Fibered knots and algebraic singularities, Topology 13 (1974), 47-59.

[Ed] Edmonds, A. L.: Deformation of maps to branched coverings in dimension two, Ann. of Math. (2) 110 (1979), 113-125

[E-N] Eisenbud, D., and Neumann, W. D.: Fibering iterated torus links, unpublished preprint (1977).

[EN] Enriques, F.: Sulla construzione delle funzioni algebriche di due variabile possedenti una data curva di diramazione, Ann. Nat. Pura Appl., 1 (1923).

[Ev] Evers, M.: Dissertation (Köln 1979).

[F-L-P] Fathi, A., Laudenbach, F., and Poénaru, V., *Travaux de Thurston sur les surfaces* Asterisque 66-67 (1979).

[Ga] Gabrielov, A.M.: Intersection Matrices for certain singularities, Engl. Translation in Funct. Anal. Appl. 7 (1973), 182-193.

[Gr] Grauert, H.: Über Modifikation und exzeptionelle analytische Mengen, Math. Ann. 146 (1962), 331-368.

[H] Hillman, J.: *Alexander ideals of links*, Springer Lect. Notes in Math. 895 (1981).

[H-N-K] Hirzebruch, F., Neumann, W.D., and Koh, S.S.: *Differentiable Manifolds and Quadratic Forms*, Lecture Notes in pure and appl. Math. 4 (Marcel Dekker, Inc. New York 1971).

[J-S] Jaco, W.H., and Shalen, P.B.: *Seifert fibered spaces in 3-manifolds*, Memoirs of the A.M.S. 220 (1979).

[Jo] Johannson, K.: *Homotopy equivalences of 3-manifolds with boundaries*, Lect. Notes in Math. 761, Springer-Verlag, New York (1979).

[K] Kähler, E.: Über die Verzweigung einer algebraischen Funktionen zweier Veränderlichen in der Umgebung einer singulären Stelle, Math. Zeit. 30 (1929), 188-206.

[K-N] Kauffman, L.H., and Neumann, W.D.: Products of knots, branched fibrations, and sums of singularities, Top. 16 (1977), 369-393.

[Kn] Kneser, H.: Glättungen von Flächenabbildungen, Math. Ann. 100 (1928), 609-617.

[Lê] Lê Dung Trang: Deux neouds algébriques FM-équivalents sont égaux, C.R. Acad. Sci., Paris, t.272 (1971), 214-216.

[Lev] Levine, J.: An algebraic classification of some knots of codimension two, Comm. Math. Helv. 45 (1970), 185-198.

[Li] Litherland, R.A.: Signatures of iterated torus knots, in *Topology of Low-dimensional manifolds, Proceedings, Sussex 1974,* Springer Lect. Notes in Math. 722 (1974), 71-84.

[Mi 1] Milnor, J.: A unique factorization theorem for 3-manifolds, Amer. J. Math. 84 (1962), 1-7.

[Mi 2] Milnor, J.: Singular points of complex hypersurfaces, Ann. of Math. Studies 61 (1968), Princeton Univ. Press.

[Mn] Montesinos, J.M.: Variedades de Seifert que son recubridores ciclicos ramificados de dos hojas, Bol. Soc. Mat. Mex. 18 (1973), 1-29.

[Mo] Morgan, J.: Nonsingular Morse-Smale flows on 3-dimensional manifolds, Topology 18 (1979), 41-53.

[Mu] Mumford, D.: The topology of normal singularities of an algebraic surface and a criterion for simplicity, Publ. Math. I.H.E.S. No. 9 (Paris 1961).

[Ne 1] Neumann, W.D.: S^1-actions and the α-invariant of their involutions, Bonner Math. Schriften 44 (Bonn 1971).

[Ne 2] Neumann, W.D.: Brieskorn complete intersections and automorphic forms, Inv. Math. 42 (1977), 285-293.

[Ne 3] Neumann, W.D.: A calculus for plumbing applied to the topology of complex surface singularities and degenerating complex curves, Trans. A.M.S. 268 (1981), 299-343.

[Ne 4] Neumann, W.D.: Invariants of plane curve singularities, in *Noeuds, Tresses et Singularités,* Monographie No. 31 de L'Enseignement Math. (1983), 223-232.

[Ne 5] Neumann, W.D.: Splicing algebraic links, Proc. U.S - Japan Seminar on Singularities 1984, Advanced Studies in Pure Math., North Holland, (to appear).

[N-R] Neumann, W.D., and Raymond, F.: Seifert manifolds, plumbing, μ-invariant, and orientation-reversing maps, *Proc. of Santa Barbara Conf.*, Springer Lect. Notes in Math. 664 (1978).

[N] Newton, I.: On the theory of fluxions and infinite series with an application to the geometry of curved lines in *The Mathematical Works of Isaac Newton,* Johnson Reprint Corporation.

[No] Northcott, D.G.: *Finite Free Resolutions*, Cambridge U. Press, Cambridge, (1978).

[O-R] Orlik, P., and Raymond, F.: Actions of SO(2) on 3-manifolds, Proc. Conf. on transformation groups, New Orleans 1967, Springer-Verlag, New York, 1968 (297-318).

[Pa] Papakyriakopolous, C.: On Dehn's Lemma and the asphericity of knots, Ann. of Math. 66 (1957), 1-26.

[Ro] Roussarie, R.: Plongements dans les variétés feuilletées et classification de feuilletages sans holonomie, Publ. Math. I.H.E.S. 43 (1974), 101-141.

[Se] Seifert H.: Topologie dreidimensionale gefaserter Räume, Acta Math. 60 (1933), 147-238.

[Si] Siebenmann, L.: On vanishing of the Rohlin invariant and nonfinitely amphicheiral homology 3-spheres, in *Topology Symposium, Siegen 1979,* Koschorke and Neumann, ed., Springer Lect. Notes 788 (1979).

[St] Stallings, J.: On fibering certain 3-manifolds, in *Topology of 3-manifolds and related topics* (Ed. by M.K. Fort, Jr.) Prentice-Hall, N.J. (1962), 95-100.

[S-W] Sumners, D.W., and Woods, J.M.: The monodromy of reducible plane curves, Inv. Math. 40 (1977), 107-141.

[Th 1] Thurston, W.P.: Three-dimensional manifolds, Kleinian groups, and hyperbolic geometry, Bull. AMS 6 (1982), 357-382 (see also *The Smith Conjecture*, eds. John W. Morgan and Hyman Bass, Academic Press Inc. (1984)).

[Th 2] Thurston, W. P.: A norm for the homology of 3-manifolds, Memoirs of the Amer. Math. Soc. (to appear).

[Th 3] Thurston, W. P.: Foliations of 3-manifolds which are circle bundles, Thesis, Berkeley 1972.

[To] Torres, G.: On the Alexander polynomial, Ann. of Math. 57 (1953), 57-89.

[vR] von Randow, R.: Zur Topologie von dreidimensionalen Baummannigfaltigkeiten, Bonner Math. Schriften 14 (Bonn 1962).

[Wa] Waldhausen, F.: On irreducible 3-manifolds that are sufficiently large, Ann. of Math. 87 (1968), 56-88.

[W] Walker, R. J.: *Algebraic Curves* Springer-Verlag, New York (1979).

[Ya] Yamamoto, M.: Classification of isolated algebraic singularities by their Alexander polynomials, Topology 23 (1984), 277-287.

[Z 1] Zariski, O.: On the topology of algebraic singularities. Amer. J. Math. 54 (1932), 453-465.

[Z 2] Zariski, O.: *Collected Papers, Volume IV: Equisingularity on Algebraic Varieties*, Ed. J. Lipman and B. Teissier (M.I.T. Press, Cambridge, Mass. 1979).

[Z-V-C] Zieschang, H., Vogt, E., and Coldewey, H.: *Surfaces and Planar Discontinuous Groups*, Lecture Notes in Mathematics 835 (Springer-Verlag, 1980).

Library of Congress Cataloging in Publication Data

Eisenbud, David, 1947-
Three-dimensional link theory and invariants of plane curve singularities.

Bibliography: p.
Includes index.
1. Link theory. 2. Invariants. 3. Curves, Plane. 4. Singularities (Mathematics)
I. Neumann, W. D. (Walter David), 1946- . II. Title. III. Title: 3-dimensional link theory and invariants of plane curve singularities.
QA612.2.E37 1985 · 514'.224 85-545
ISBN 0-691-08380-0
ISBN 0-691-08381-9 (pbk.)

David Eisenbud is Professor of Mathematics at Brandeis University. Walter Neumann is Professor of Mathematics at the University of Maryland.

Milton Keynes UK
Ingram Content Group UK Ltd.
UKHW010153100124
435723UK00001B/16

9 780691 083810